Solutions Manual to Accompany

CONVECTIVE HEAT AND MASS TRANSFER

THIRD EDITION

W. M. KAYS
STAMFORD UNIVERSITY

M.E. CRAWFORD
THE UNIVERSITY OF TEXAS AT AUSTIN

McGraw- Hill, Inc.
New York St. Louis San Francisco Auckland Bogotá
Caracas Lisbon London Madrid Mexico Milan Montreal
New Delhi Paris San Juan Singapore Sydney Tokyo Toronto

Solutions Manual to Accompany Kays/Crawford
CONVECTIVE HEAT AND MASS TRANSFER
Third Edition

ISBN 0-07-033722-5

1234567890 BKM BKM 90987654

Solutions Manual to accompany

CONVECTIVE HEAT AND MASS TRANSFER

Third Edition

W. M. Kays
Emeritus Professor of Mechanical Engineering
Stanford University

M.E. Crawford
Professor of Mechanical Engineering
The University of Texas at Austin

An Introductory Note

Some of the problems in the text are brief exercises leading to single numerical or algebraic results, but the great majority are much more extensive investigations, some approaching the magnitude of term projects. In the latter cases, there is usually no simple answer. Student initiative is encouraged and this leads to results that may differ numerically or may involve results not asked for in the problem statement. In any case, the authors place more value on a written discussion at the end of the student's papers, and on the development of the analysis, than on numerical results.

It is not practicable to provide a "solutions manual" containing examples of complete papers for assignments of this kind. The authors have chosen rather to provide, in a very abbreviated form, some of the key results for these problems. In some cases rather than give numerical results, a brief discussion of how to attack the problem is provided. Only a small fraction of the problems can be used in any one course, and it is hoped that instructors will find a sufficient number of problems to satisfy a variety of needs, including differing tastes and interests, and differing teaching styles.

Chapter 4

4-1

Let $x_1 = x$, $x_2 = y$, $u_1 = u$, and $u_2 = v$ in Eq. 4-17. Then:

Equation, x - direction: $$\frac{\partial P}{\partial x} = \frac{\partial^2 u}{\partial x^2}$$

Equation, y - direction: $$\frac{\partial P}{\partial y} = 0$$

Thus, $P = P(x)$ and $\frac{\partial P}{\partial x} = \frac{dP}{dx}$. The equation becomes

$$\frac{dP}{dx} = \mu \frac{d^2 u}{d y^2}$$

4-2

For 2- dimensional flow, with $x_1 = x$, $x_2 = y$, $u_1 = u$, and $u_2 = v$ in Eq. (4-17),

x - dir. $$0 + \frac{\partial P}{\partial x} = \mu \left(\frac{\partial^2 u}{\partial x^2} + \frac{\partial^2 u}{\partial y^2} \right)$$

y - dir.: $$0 + \frac{\partial P}{\partial y} = \mu \left(\frac{\partial^2 v}{\partial x^2} + \frac{\partial^2 v}{\partial y^2} \right)$$

Differentiate and add these two equations:

$$\nabla^2 P = \mu \left[\frac{\partial}{\partial x} \left(\frac{\partial^2 u}{\partial x^2} + \frac{\partial^2 u}{\partial y^2} \right) + \frac{\partial}{\partial y} \left(\frac{\partial^2 v}{\partial x^2} + \frac{\partial^2 v}{\partial y^2} \right) \right]$$

$$\nabla^2 P = \mu \left[\frac{\partial^2}{\partial x^2} \left(\frac{\partial u}{\partial x} + \frac{\partial v}{\partial y} \right) + \frac{\partial^2}{\partial y^2} \left(\frac{\partial u}{\partial x} + \frac{\partial v}{\partial y} \right) \right]$$

$$\nabla^2 P = 0$$

4-3

From (4-11) or (4-12), and using the fact that along the centerline $\partial u / \partial r = 0$,

$$\rho u \frac{\partial u_{cl}}{\partial x} + \frac{dP}{dx} = 0$$

or,
$$\frac{\partial}{\partial x}\left(\frac{\rho u_{cl}^2}{2}\right)+\frac{dP}{dx}=0$$

or,
$$P+\rho\frac{u_{cl}^2}{2}=\text{const}$$

(Similar to external b.l. flow outside of the b.l.)

4-4

Consider the continuity equation (4-5) in cylindrical coordinates with $\nabla \cdot \mathbf{V}$ given on page 567. Let **r = R + y** and **z = Rϕ**, where **R** is the radius of the inner surface. Note that $\partial/\partial x = 0$ (no axial flow). Thus

$$R\frac{\partial v_\varphi}{\partial z}+R\frac{\partial v_r}{\partial y}+v_r+y\frac{\partial v_r}{\partial y}=0$$

For v_r small and **y << R**, the last two terms in the equation are negligible compared to the first two. Thus the equation becomes identical to (4-7) in Cartesian coordinates. (Actually these are curvilinear coordinates with **y << R**. Using similar arguments, the applicable momentum equation is (4-11), with the convective acceleration terms neglected. The applicable energy equation is (4-38).

4-5

The appropriate set of coordinates is the fixed, non-rotating cylindrical system with **r**, the radial direction; ϕ, the circumferential direction; and **x**, the axial direction above the disk. In the governing equations, all three velocity components will appear, but derivatives with respect to ϕ will be zero due to rotational symmetry.

The flow over a rotating disk is boundary layer in character, but the complete Navier-Stokes equations can be solved in exact form (see Schlichting, ref. 2, page 93). The applicable energy equation is (4-35) with the conduction gradient in the r-direction neglected. (See Schlichting, page 296, for references to heat transfer solutions.)

4-6

The applicable energy equation is (4-31). Assume no mass diffusion. From the definition of enthalpy, **i = e + P/ρ**, substitute into (4-31) and let ρ = **const**.

$$\rho\frac{De}{D\theta}-\nabla\cdot k\nabla t-S=0$$

For no fluid motion,

$$\rho\frac{\partial e}{\partial\theta}-\nabla\cdot k\nabla t-S=0$$

This is the classic heat conduction equation for a solid with **de = cdt.** For steady conduction and constant, Poisson's equation is obtained:

$$\nabla^2 t-\frac{S}{k}=0$$

4-7

Eq. (4-1) can be considered a continuity equation for the **j** - component of a mixture, but the creation term must be added:

$$\frac{\partial G_{tot,j,x}}{\partial x} + \frac{\partial G_{tot,j,y}}{\partial y} + \frac{\partial \rho_j}{\partial \theta} = \dot{m}_j'''$$

Now substitute $G_{tot,j} = G_{diff,j} + G_{conv,j}$ and consider steady flow. Next substitute $G_{conv,j} = m_j G$ and $G_{diff,j} = -\gamma_j \nabla m_j$. (Recall that **G** is the total mass flux vector.) Ignore $G_{diff,j,x}$ as a boundary layer approximation.

$$m_j\left[\frac{\partial G_x}{\partial x} + \frac{\partial G_y}{\partial y}\right] + G_x\frac{\partial m_j}{\partial x} + G_y\frac{\partial m_j}{\partial y} - \frac{\partial}{\partial y}\left(\gamma_j \frac{\partial m_j}{\partial y}\right) = \dot{m}_j'''$$

4-8

For conservation of mass, the mass flow rates, inflow on the radial face and the axial face, are:

$$\dot{m}_r = G_r A_r \qquad G_r = \rho v_r \,;\; A_r = r\,d\phi\,dx$$

$$\dot{m}_x = G_x A_x \qquad G_x = \rho u \,;\; A_x = r\,d\phi\,dr$$

Application of the procedure leading up to equation (4-2), assuming steady flow, yields:

$$\frac{\partial(\dot{m}_x)}{\partial x} + \frac{\partial(\dot{m}_r)}{\partial r} = 0 \qquad \frac{\partial(\rho u)}{\partial x} + \frac{1}{r}\frac{\partial(r\rho v_r)}{\partial r} = 0$$

Note the **1/r** term comes from dividing through the equation by the differential volume, **(rdϕ dr dx)**. For constant properties, Eq. (4-9) is found:

$$\frac{\partial(u)}{\partial x} + \frac{1}{r}\frac{\partial(r v_r)}{\partial r} = 0$$

For x-momentum , the x-momentum flow rates, inflow on the radial face and the axial face, and the x-forces are:

$$\dot{M}_r = u G_r A_r$$

$$\dot{M}_x = u G_x A_x$$

$$F_x = -\sigma_x A_x - \tau_{rx} A_r \qquad \sigma_x \approx -P \qquad \tau_{rx} = \mu\frac{\partial u}{\partial r}$$

Application of the procedure leading up to Eqs. (4-10) and (4-11), assuming steady flow, yields, for constant properties:

$$\rho u\frac{\partial u}{\partial x} + \rho v_r\frac{\partial u}{\partial r} + \frac{dP}{dx} = \mu\frac{1}{r}\frac{\partial}{\partial r}\left(r\frac{\partial u}{\partial r}\right)$$

For energy , the energy flow rates, inflow on the radial face and the axial face, and the x-forces are:

$$\dot{E}_{conv,r} = G_r\left(i+\tfrac{1}{2}u^2\right)A_r \qquad \dot{E}_{conv,x} = G_x\left(i+\tfrac{1}{2}u^2\right)A_x$$

$$\dot{q}_r = -kA_r\frac{\partial t}{\partial r} \qquad \dot{q}_x = -kA_x\frac{\partial t}{\partial x}$$

$$\dot{W}_{shear,r} = u\tau_{rx}A_r$$

Application of the procedure leading up to Eqs. (4-27) and (4-28), assuming steady flow and no mass diffusion, yields, for constant properties:

$$\rho u\frac{\partial i}{\partial x} + \rho v_r\frac{\partial i}{\partial r} - k\frac{1}{r}\frac{\partial}{\partial r}\left(r\frac{\partial t}{\partial r}\right) - k\frac{\partial}{\partial x}\left(\frac{\partial t}{\partial x}\right) - \mu\left(\frac{\partial u}{\partial r}\right)^2 - u\frac{dP}{dx} = 0$$

or, for ideal gases and incompressible liquids,

$$u\frac{\partial t}{\partial x} + \rho v_r\frac{\partial t}{\partial r} - \alpha\frac{1}{r}\frac{\partial}{\partial r}\left(r\frac{\partial t}{\partial r}\right) - \alpha\frac{\partial}{\partial x}\left(\frac{\partial t}{\partial x}\right) - \frac{Pr}{c}\left(\frac{\partial u}{\partial r}\right)^2 - \frac{1}{\rho c}u\frac{dP}{dx} = 0$$

4-9

These steps are essentially a repeat of the formulation leading up to Eq. (4-39).

Chapter 5

5-1

Note that Eqs. (5-9) and (4-17) are both unsteady forms of the Navier-Stokes equations written in index notation, although (5-9) is for compressible flow with variable viscosity, whereas (4-17) is for constant density flow with constant viscosity. Their forms differ in the convection term and in the stress (or diffusion) term. Chain-rule differentiate the convective term of (5-9) and subtract (5-8) to modify the convection term, then assume constant density for the time term. This recovers the left hand side of (4-17). Now, substitute the stress tensor, (5-6) into (5-9) and substitute (5-3) for the strain-rate tensor. Assume constant viscosity and constant density. For constant density flow, the second coefficient of viscosity (the dilatation term) is eliminated. The second term in the strain rate tensor also disappears from conservation of mass (assuming constant density), and the result is the right hand side of (4-17).

5-2

Note that Eq. (5-11) is a stagnation enthalpy equation, whereas (4-32) is a static enthalpy equation. Multiply the Eq. (5-10) by **u** and subtract it from (5-11) (you will first have to chain-rule the viscous work term in 5-11) to obtain the unsteady static enthalpy equation. Now, substitute (5-6) along with the strain-rate tensor (5-3) into the static enthalpy equation. The result is (4-32).

5-3

let:

$$i^* = i + \tfrac{1}{2} u_i u_i$$

and substitute decompositions for $\mathbf{i}^*$ and $\mathbf{u_i}$

$$\overline{i^*} + i^{*\prime} = \bar{i} + i' + \tfrac{1}{2}(\bar{u}_i + u_i')(\bar{u}_i + u_i')$$

Now, expand the velocity term:

$$\overline{i^*} + i^{*\prime} = \bar{i} + i' + \tfrac{1}{2}\bar{u}_i \bar{u}_i + (\bar{u}_i u_i') + \tfrac{1}{2}(u_i' u_i')$$

Now, collect the time-averaged terms and the fluctuating terms. The result is (5-16), as can be found in Ref. 1.

5-4

These steps are essentially a repeat of the formulation leading up to Eq. (5-21).

5-5

The first step is to chain-rule the convective term on the right-hand-side of the equation and apply the conservation of mass equation (5-22). Then apply the index notation rule that repeated subscripts sum, **j** = 1, 2. Note there will not be a **j** = 3 term because of the two-dimensional boundary layer assumption (**w** = 0). The pressure gradient becomes an ordinary derivative. The diffusion operator changes under the boundary layer assumptions to

$$\frac{\partial(\ )}{\partial x_j} \rightarrow \frac{\partial(\ )}{\partial y}$$

In the viscous stress tensor, τ_{ji}, only one term will be present, due to the boundary layer approximation and the constant density assumption. Likewise for the Reynolds stress term. The result is:

$$\bar{\tau}_{ji} - \rho\overline{u_j' u_i'} \rightarrow \bar{\tau}_{yx} - \rho\overline{v' u'}$$

This stress is τ (= τ_{yx}) in Eq. (5-27). Now divide through by the density, and the constant density assumption allows the density to be moved into the diffusion term; μ becomes ν, and the density is removed from the Reynolds stress. Note Eq. (5-28) remains valid for variable viscosity, but not for variable density.

5-6

These steps are essentially a repeat of the formulation leading up to Eq. (5-31), followed by the formulation leading to Eq. (5-34). Eq. (5-34) is valid for either an ideal gas or an incompressible liquid (both obeying **de = c dt**), and it is valid for variable thermal conductivity, but not for variable density.

5-7

These steps are essentially a repeat of the formulation leading up to Eq. (5-38).

5-8

$$\bar{u}\frac{\partial \bar{u}}{\partial x} + \bar{v}\frac{\partial \bar{u}}{\partial y} = \frac{\partial}{\partial y}\left(\nu\frac{\partial \bar{u}}{\partial y} - \overline{u'v'}\right) - \frac{d\bar{P}}{dx}$$

$$\bar{u}\frac{\partial \bar{t}}{\partial x} + \bar{v}\frac{\partial \bar{t}}{\partial y} = \frac{\partial}{\partial y}\left(\alpha\frac{\partial \bar{t}}{\partial y} - \overline{v't'}\right)$$

Convection( ) = Diffusion( ) ± Source( )

Both the momentum equation and energy equations have a convective term and a diffusive term, but only the momentum equation has a source term. An equivalent energy source term would be a viscous dissipation term, which could be added.

Chapter 6

6-1

Let **r** = **R** be the wall and **r** = **y** be the edge of the control volume. Typical terms would be:

$$M_x = 2\pi \int_{R-y}^{R} \rho u^2 r dr \quad \text{and} \quad M_y = (R-y)2\pi\delta x\, v_y \rho_y u_y$$

$$\text{Normal stress} = P\int_{R-y}^{R} 2\pi r dr$$

$$\text{Shear stress:} \qquad \tau_0 2\pi R\delta x; \qquad \tau_y 2\pi(R-y)\delta x$$

Now follow the procedure on page 65:

$$-\tau_0 = \frac{1}{R}\frac{d}{dx}\left[\int_{R-y}^{R} \rho u^2 r dr\right] - \frac{u_\infty}{R}\frac{d}{dx}\left[\int_{R-y}^{R} \rho u r dr\right] - \rho_\infty u_\infty y\left(1-\frac{y}{2R}\right)\frac{du_\infty}{dx}$$

6-2

Let **r** = **R** be the wall and **r** = **y** be the edge of the control volume. Typical terms would be:

$$\dot{E}_{conv,x} = 2\pi \int_{R-y}^{R} \rho u i_s r dr$$

$$\dot{q}_R = 2\pi R\delta x\, \dot{q}_R''$$

Now follow the procedure on pages 69-70:

$$-\dot{q}_R'' = \frac{1}{R}\frac{d}{dx}\int_{R-y}^{R} \rho u i_s r dr$$

6-3

Assume: Fick's Law holds; steady flow; concentration boundary layer ≥ momentum boundary layer thickness; y-direction gradients >> x-direction gradients; **G** at y = 0 is in y-direction. Let $m_{j,o}$ be the mass concentration of j at the wall. For the j-component, typical terms are:

$$\text{Inflow of } j = \int_0^y G_x m_j dy + G_{y,0} m_{j,0}\delta x - \gamma_j\left(\frac{dm_j}{dy}\right)_0 \delta x$$

$$\text{Rate of creation of j} = \int_0^y \dot{m}_j''' \delta x \, dy$$

The conservation principle that Outflow - Inflow + Increase in Storage = Rate of Creation for specie **j** is used to obtain the integral equation. Continuity Eq. (6-1) is used and $m_{j,y} = m_{j,\infty}$. Then,

$$-\gamma_j \left(\frac{\partial m_j}{\partial y} \right)_0 = \frac{d}{dx} \left[\int_0^y \rho u (m_j - m_{j,\infty}) dy \right]$$

$$-\rho_0 v_0 (m_{j,0} - m_{j,\infty}) - \int_0^y \dot{m}_j''' dy$$

6-4

These steps are essentially a repeat of the formulation leading up to Eq. (6-8) and Eq. (6-21).

Chapter 7

7-1

Let **2d** be the plate spacing. Then the entry length, **L**, is calculated to be:

$$\frac{L}{2d} = \frac{Re_{2d}}{38.6} \qquad \text{where} \qquad Re_{2d} = \frac{V(2d)}{\nu}$$

The assumed velocity profile is:

$$V = u_\infty\left(1 - \frac{\delta}{3d}\right)$$

d is related to axial distance **x** by:

7-2

$$\frac{15\nu}{Vd}\frac{x}{d} = 10.5\frac{\delta}{d} - 40.5 + \frac{40.5}{\left(1 - \frac{\delta}{3d}\right)} + 72\ln\left(1 - \frac{\delta}{3d}\right)$$

For constant properties:

$$\Delta P = C_f \frac{A}{A_c} \rho \frac{V^2}{2}$$

If density is variable:

$$\Delta P = \frac{G^2}{2}\left[2\left(\frac{1}{\rho_2} - \frac{1}{\rho_1}\right) + C_f \frac{A}{A_c}\frac{1}{\rho_m}\right]$$

$$\text{where} \quad \rho_m = \frac{1}{\frac{1}{A}\int_0^A \frac{1}{\rho}\,dA}$$

(Note: $\frac{r_h}{L} = \frac{A_c}{A}$)

7-3

The actual momentum flux divided by the momentum flux from the mean velocity is 1.33.

7-4

Let the plate spacing be **d**, and measure **y** from the centerline. Then,

$$u = \left(\frac{d^2}{8\mu}\right)\left(-\frac{dP}{dx}\right)\left(1 - \frac{4y^2}{d^2}\right)$$

$$V = \left(\frac{d^2}{12\mu}\right)\left(-\frac{dP}{dx}\right)$$

$$u = \frac{3}{2}V\left(1 - \frac{4y^2}{d^2}\right)$$

$$C_f = \frac{24}{Re} \qquad \text{where } Re = 2dVr/\mu$$

7-5

The pressure difference is **14.8 Pa.**

The Reynolds number in the larger tube is **865**; in the smaller **108**.

7-6

A 10 percent displacement if the partition (fins) between two passages results in a 77 percent difference in the mass flow rates, the higher flow rate in the larger passage. The total flow rate for the two passages with unequal fin spacing will be 2.7 percent greater than for the same two passages with the separating fin equidistant from adjacent fins.

7-7

These steps are essentially a repeat of the formulation leading up to Eq. (7-12).

7-8

(see problem 7-4, basically a duplicate as stated)

7-9

Let the annulus be defined by inner radius, r_i , and outer radius, r_o , and radius ratio, $r_i/r_o = r^*$. From Eq. (4-11), for constant properties, the fully-developed x-momentum equation is:

$$\frac{dP}{dx} = \mu\frac{1}{r}\frac{\partial}{\partial r}\left(r\frac{\partial u}{\partial r}\right)$$

Integration of the equation and application of the no-slip boundary conditions at r_i and r_o yields:

$$u = C\left[1 - \left(\frac{r}{r_o}\right)^2 + B\ln\left(\frac{r}{r_o}\right)\right]$$

$$B = \frac{\left(r^{*2} - 1\right)}{\ln\left(r^*\right)} \qquad C = \frac{1}{\mu}\left(\frac{-dP}{dx}\right)\frac{r_o^2}{4}$$

$$V = \frac{C}{2}\left(1 + r^{*2} - B\right)$$

The resulting friction coefficients for the i- and o-surfaces are:

$$\tau_o = \mu \left.\frac{\partial u}{\partial r}\right|_{r_o} \qquad C_{f,o} = \frac{2\tau_o}{\rho V^2} = \frac{4\mu}{\rho V r_o}\left(\frac{B-2}{1+r^{*2}-B}\right)$$

$$\tau_i = \mu \left.\frac{\partial u}{\partial r}\right|_{r_i} \qquad C_{f,i} = \frac{2\tau_i}{\rho V^2} = \frac{4\mu}{\rho V r_o}\left(\frac{\frac{B}{r^*}-2r^*}{1+r^{*2}-B}\right)$$

Define the Reynolds number based on the hydraulic diameter, $D_h = 2(r_o - r_i)$, and define a perimeter-averaged friction coefficient,

$$C_f = \frac{C_{f,i}\, r_i + C_{f,o}\, r_o}{r_i + r_o} \qquad Re = \frac{2 r_o (1-r^*)\rho V}{\mu}$$

$$C_f\, Re = \left(\frac{C_{f,i}\, r^* Re + C_{f,o}\, Re}{r^*+1}\right) = \frac{16(1-r^*)^2}{1+r^{*2}-B}$$

Using these definitions, Figure 7-5 (and Shah and London, Table 86) can be reproduced.

7-10

Start with Eq. (7-18) and rearrange the convective terms by adding the conservation of mass Eq. (4-18), assuming steady flow.

$$\frac{\mu}{r}\frac{\partial}{\partial r}\left(r\frac{\partial u}{\partial r}\right) = \rho u\frac{\partial u}{\partial x} + \rho v_r\frac{\partial u}{\partial r} + \frac{dP}{dx}$$

or,

$$0 = \frac{\partial(\rho u u)}{\partial x} + \frac{1}{r}\frac{\partial(r\rho v_r u)}{\partial r} + \frac{dP}{dx} - \frac{1}{r}\frac{\partial}{\partial r}(r\tau_{rx})$$

Now, apply the integral operator:

$$\int_0^x dx\int_0^{r_o} 2\pi r\,dr : \left[0 = \frac{\partial(\rho u u)}{\partial x} + \frac{1}{r}\frac{\partial(r\rho v_r u)}{\partial r} + \frac{dP}{dx} - \frac{1}{r}\frac{\partial}{\partial r}(r\tau_{rx})\right]$$

or, for the limits **x** = 0 (entrance) to **x**, where $v_r = 0$ at $r = 0$ and at r_o, and where $\tau_{rx} = 0$ to $r = r_o$ where $\tau_{rx} = \tau_o$,

$$0 = \left[2\pi\int_0^{r_o}(\rho u u)r\,dr - 2\pi\frac{r_o^2}{2}\rho V V\right] + [0-0] + 2\pi\frac{r_o^2}{2}\int_0^x \frac{dP}{dx}dx - \left[2\pi r_o\int_0^x \tau_o\,dx - 0\right]$$

or,

$$\Delta P A_c = (P_0 - P_x)A_c = \int_{A_c}(\rho u^2)dA_c - \rho V^2 A_c + \frac{A_c}{D_h/4}\int_0^x \tau_o\,dx$$

Divide by A_c and factor out $\frac{1}{2}\rho V^2$, and define the mean friction coefficient:

$$c_{f_m} = \frac{1}{x}\int_0^x c_f\,dx = \frac{1}{x}\int_0^x \frac{\tau_o}{\frac{1}{2}\rho V^2}dx$$

For constant density,

$$\frac{\Delta P}{\frac{1}{2}\rho V^2} = \frac{2}{A_c}\int_{A_c}\left(\frac{u}{V}\right)^2 dA_c - 2 + \frac{x}{D_h/4}c_{f_m}$$

In this expression the first two terms of ΔP represent the change in pressure due to acceleration of the flow, while the last term represents the frictional pressure drop component. Comparing this result to Eq. (7-19) provides the definition of the apparent friction factor.

$$\bar{c}_{f_{app}} = c_{f_m} + \frac{D_h/4}{x}\left[\frac{2}{A_c}\int_{A_c}\left(\frac{u}{V}\right)^2 dA_c - 2\right]$$

7-11

Set up the problem for a value of **Re**, say 1000. Choose properties for any fluid; since this is a constant-property problem, the choice will not matter. The integration will be carried out from x = 0 to **x** > entry length, as determined by $\mathbf{Re/(x/D_h)} \approx 10$ to assure fully-developed flow. If you choose air at room temperature, then a pipe radius 4 cm and a length of 8 m should be sufficient.

To verify Figure 7-7 the pressure drop must be determined from the numerical calculation, and = $\bar{c}_{f_{app}}$ computed from its definition, Eq. (7-19). Note the definition of fully-developed flow is quite arbitrary due to the asymptotic behavior of $\bar{c}_{f_{app}}$ with $\mathbf{Re/(x/D_h)}$.

Plotting the ratio $\mathbf{u_{cl}/V}$ shows a continual acceleration of the centerline velocity as the velocity profile changes shape, over the limits $1 \le \mathbf{u_{cl}/V} \le 2$, which occurs at about

$\mathbf{Re/(x/D_h)}$ = 5-10. This ratio is a measure of fully-developed flow. Note, the wall friction will become fairly constant by $\mathbf{Re/(x/D_h)}$ = 20, but $\bar{c}_{f_{app}}$ continues to change until $\mathbf{Re/(x/D_h)} \approx 10$. It is worthwhile to plot a set of normalized velocity profiles, **u/V**, over $0 \le \mathbf{r/r_o} \le 1$ to show how the initial flat profile changes (see Figure 7-6).

7-12

Set up the problem for a value of **Re**, say 1000. Choose properties for any fluid; since this is a constant-property problem, the choice will not matter. The integration will be carried out from **x** = 0 to **x** > entry length, as determined by $\mathbf{Re/(x/D_h)} \approx \mathbf{20}$ to assure fully-developed flow for both parallel planes flow and annular flow. Be sure to use the hydraulic diameter **Re**. If you choose air at room temperature, then for the parallel planes, a channel width of 2 cm and a length of 4 m should be sufficient. For the annulus at this small $\mathbf{r^*}$ the length should be about 40 m.

The comments of Problem 7-11 apply to this problem. You will find that $\bar{c}_{f_{app}}$ for the pipe, parallel planes, and annulus plot identically for $\mathbf{Re/(x/D_h)} \ge 50$. At 50, the solutions diverge somewhat, to approach their respective $\bar{c}_{f_{app}}$ **Re** asymptotes. Be careful in you interpretation of $\bar{c}_{f_{app}}$ **Re**, as the apparent friction coefficient contains the entry length effect, whereas the fully-developed friction coefficient does not. Compare your fully-developed $\mathbf{c_f}$ **Re** to Figure 7-5. You can compute $\mathbf{c_{f,i}}$ **Re** and compare it to Shah and London (or solve Problem 7-9); likewise for $\mathbf{c_{f,o}}$ **Re**. You will find $\mathbf{c_{f,i}}$ **Re** varies strongly with $\mathbf{r^*}$ whereas $\mathbf{c_{f,o}}$ **Re** is a weak function., and their combination, $\mathbf{c_f}$ **Re**, being perimeter- averaged, varies between the pipe limit and the parallel planes limit. By $\mathbf{r^*}$ of between 0.3 and 0.6, the parallel plane behavior is achieved.

You will find the limits $1 \le \mathbf{u_{cl}/V} \le 1.5$ for the parallel planes will be only slightly different for the annulus (1.56). You can plot the evolution of the velocity profiles for the parallel planes case over the distance from wall to centerline. For the annulus, you need to plot from $\mathbf{r_i}$ to $\mathbf{r_o}$. For $\mathbf{r^*} = 0.1$, you will see the slight offset of $\mathbf{u_{max}}$ from being at the arithmetic average of the inner and outer radius.

Chapter 8

8-1

The stagnation point flow has a similarity solution since $u_\infty = Cx^m$, where $C = 4V/D$, $m = 1$. Using similarity variables, and the definition of displacement thickness,

$$\delta_1 = \sqrt{\frac{\nu}{C}}\int_0^\infty [1-\varsigma'(\eta)] = \sqrt{\frac{\nu}{C}} \lim_{\eta\to\infty} [\eta-\varsigma(\eta)]$$

For $\eta \geq 3, [\eta - \varsigma(\eta)] \to 0.648$. Thus δ_1 = **0.0955** mm.

8-2

In complex variable theory, potential flow solutions are greatly simplified if the flow domain can be transformed into a semi-infinite domain. For uniform parallel flow the complex potential function is $F(z) = u_\infty z$. (Recall that $dF/dz = \overline{w} = u - iv$). A conformal transformation is now proposed such that $F(z) = Aw(z)$. For a wedge of opening β, the transformation to the upper half of the w-plane occurs when

$$F(z) = Aw(z) = Az^{\pi/8} \quad \text{and } \gamma = \pi - \beta/2$$

Now $$|\overline{w}| = |A\frac{dw}{dz}| = \frac{A\pi}{\gamma} z^{\frac{\pi}{\gamma}-1}$$

Let $$m = \frac{\pi}{\gamma} - 1 = \frac{\pi}{\pi - \beta/2} - 1 \quad \text{or} \quad m = \frac{\beta/\pi}{2-\beta/\pi}$$

Check to see if this gives $u_\infty = Cx^m$. Let $z = re^{i\phi}$ and $r = x$, $\phi = \gamma$.

Thus $$|\overline{w}| = u_\infty = |\frac{A\pi}{\lambda} x^m e^{i\gamma m}| = Cx^m$$

8-3

The applicable equation is (8-31). Substitute the given velocity profile into the wall gradient term (LHS). For the RHS use Eq. (6-6) to obtain $d\delta_2/dx$ in terms of $d\delta/dx$.. Separate variable in (8-31) and integrate to obtain $\delta/x = 4.78\sqrt{Re_x}$. δ_1, δ_2, and c_f follow from (6-5, 6, 10), along with the results for $\delta(x)$.

$$\frac{\delta_1}{x} = 1.735 Re_x^{-1/2} \qquad \frac{\delta_2}{x} = 0.647 Re_x^{-1/2}$$

$$C_f = 0.657 Re_x^{-1/2}$$

8-4

The applicable momentum integral equation is (6-7). With the density term, Eq.(8-40) has an extra term on the LHS: $\dfrac{2\delta_2^2}{\nu}\dfrac{u_\infty}{\rho}\dfrac{d\rho}{dx}$

Following the procedure outlined on Page 104,

$$d\left(R^2\rho^2\delta_2^2u_\infty^b\right) = a\nu u_\infty^{b-1}R^2\rho^2\,dx$$

where $a = 0.44$, $b = 5.68$

8-5

Assume $\delta_1 = 0$ to determine $\dot{m}$ and $V(x{=}0)$. Use (8-42) to find δ_2 and Table 8-4, with λ being computed, to find H and δ_1. Then,

$$\delta_2 = \frac{0.664\,\nu^{1/2}}{R_{x=L}\,V_{x=L}^{2.84}}\left[\int_0^{\pi/2}\left(V_a + \frac{V_b - V_a}{L}r\theta\right)^{4.86}(0.05 - r\sin\theta)^2\,r\,d\theta\right]^{1/2}$$

Results: $0 \le \delta_1 \le 0.05\text{m}, \delta_2 = 0.0127\text{cm}$

$\Delta P = 60\text{Pa}$ $\quad\lambda = 0.17$ $\quad H = 2.1$

$\delta_1 = 0.0267$ cm

$$\dot{m} = \rho V_{x=L}\left[\frac{\pi}{4}\left(\frac{2.5 - 2\delta_1}{100}\right)^2\right] = 0.005642\ \text{kg/s}$$

$$C = \frac{\dot{m}_{actual}}{\dot{m}_{\delta_1=0}} = 0.956$$

Note: $Re\uparrow,\ \delta_2\downarrow,\ \therefore C \to 1.0$

8-6

Start with Eq. (8-23), and introduce the stream function, Eq. (8-9), to transform the dependent variables **u** and **v**.

$$\nu\frac{\partial^3\psi}{\partial y^3} = \frac{\partial\psi}{\partial y}\frac{\partial}{\partial x}\left(\frac{\partial\psi}{\partial y}\right) - \frac{\partial\psi}{\partial x}\left(\frac{\partial^2\psi}{\partial y^2}\right) - \frac{u_\infty^2 m}{x}$$

Now, transform the independent variables, **x** and **y** using the Blasius variable Eq. (8-11) for **y,** and define the stream function using Eq. (8-10):

$$\nu\frac{\partial^3\psi}{\partial y^3} = \frac{\partial\psi}{\partial y}\frac{\partial}{\partial x}\left(\frac{\partial\psi}{\partial y}\right) - \frac{\partial\psi}{\partial x}\left(\frac{\partial^2\psi}{\partial y^2}\right) - \frac{u_\infty^2 m}{x}$$

$$\xi = x$$

$$\eta = y\sqrt{\frac{u_\infty}{\upsilon x}} = y\sqrt{\frac{C}{\upsilon}}\,x^{(m-1)/2}$$

$$\psi(\xi,\eta) = G(\xi)\varsigma(\eta) = \sqrt{\upsilon x u_\infty}\,\varsigma(\eta) = \sqrt{\upsilon C}\,x^{(m-1)/2}\,\varsigma$$

Transform the various derivatives using the chain rule:

$$\frac{\partial(\,)}{\partial x} = \frac{\partial(\,)}{\partial\xi}\frac{\partial\xi}{\partial x} + \frac{\partial(\,)}{\partial\eta}\frac{\partial\eta}{\partial x}$$

$$\frac{\partial(\,)}{\partial y} = \frac{\partial(\,)}{\partial\xi}\frac{\partial\xi}{\partial y} + \frac{\partial(\,)}{\partial\eta}\frac{\partial\eta}{\partial y}$$

At this point, you can do one of two procedures: the first is to transform the ψ equation differential operators, keeping ψ as the independent variable, and then introduce $\psi(G,\zeta)$ to obtain an equation in **G** and ζ. Upon introducing the functional form of **G**, the equation separate separates and Eq. (8-24) is obtained. A second procedure is to transform the differential operators and simultaneously, substitute for $\psi(G,\zeta)$. Either works.

8-7

Set up the problem for laminar flow over a flat plate with free stream velocity of 15 m/s and free stream temperature of 300 K; let the fluid be air at 1 atm; plate length is 0.2 m; constant properties; start calculations at an **x**-Reynolds number of about 1000. Evaluate the initial velocity profile by converting the Table 8-1 values from their similarity form to physical variables.

The computations should go out to a $\mathbf{Re_x}$ of about 2 x 10^5, and the $\mathbf{Re_{\delta 2}}$ should range from an initial value of about 20 out to about 290. The shape factor should remain at about 2.59. Compare the boundary layer thickness, δ, growth (location where $\mathbf{u/u_\infty}$ is 0.99) to Eq. (8-33); compare the displacement thickness, δ_1, growth to Eq. (8-17); and compare the growth of the momentum thickness, δ_2, to Eq. (8-18). Convert several velocity profiles from physical variables to ζ' versus η to verify that similarity exists for various **x**-locations, and compare to Table 8-1. Log-log plot the friction coefficient, $\mathbf{c_f}$, versus $\mathbf{Re_x}$ to compare with Eq. (8-13), and versus $\mathbf{Re_{\delta 2}}$ to compare with Eq. (8-16). From the momentum thickness distribution, create its x-derivative and compare the results to Eq. (6-11).

8-8

Set up the problem for laminar flow over a flat plate with free stream velocity varying as **Cx^m** (m/s) and free stream temperature of 300 K; let the fluid be air at 1 atm; plate length is 0.01 m; constant properties.

For m=0.111, start the calculations at an **x**-Reynolds number of about 1000. For **Pr** = 0.7 and 1.0, let the fluid be air at 1 atm, and for 5.0, let the fluid be water. Stop the calculations at an **x**-Reynolds number of about 200,000. For this **m** value, **C** = 592 yields a starting location of 0.0001 m and an ending location of 0.012 m. For **m**=1.0, the same **C** yields a starting **x**-Reynolds number of about 100 at an x value of 0.002 m and an ending location of 0.12 m yields an ending **x**-Reynolds number of about 300,000. Determination of the starting profile is a bit difficult, as the text provides no table entries other than for m = 0. These values can typically be found in fluid mechanics texts.

The computations will have limited comparisons, but you should observe the sqrt(**x**) variation of the variables, and the friction coefficient can be compared to Eq. (8-25) along with Table 8-2. For **m**=0.111, the shape factor will be 2.47 and for **m**=1.0, the value will be 2.21. The friction coefficient can be computed using the momentum integral Eq. 6-8. Similarity can be confirmed by plotting velocity profiles at different **x**-locations.

Chapter 9

9-1

Let the plate spacing be **d**. Let **y** = 0 be on plate no. 1. Then the temperature profile for any heat flux ratio will be:

$$\frac{t - t_1}{t_m - t_1} = \frac{\left[\frac{1}{2}\frac{y^4}{d^4} - \frac{y^3}{d^3} + \frac{y}{d\left(1 + \dot{q}_2''/\dot{q}_1''\right)}\right]}{\left[\frac{1}{2}\frac{1}{\left(1 + \dot{q}_2''/\dot{q}_1''\right)} - \frac{9}{70}\right]}$$

The mean temperature is:

$$t_m = \frac{V}{\alpha}\left(\frac{dt_m}{dx}\right)d^2\left(\frac{9}{70} - \frac{1}{2\left[1 + \dot{q}_2''/\dot{q}_1''\right]}\right) + t_1$$

$$\left(\frac{dt_m}{dx}\right) = \frac{\dot{q}_1'' + \dot{q}_2''}{V\rho d c}$$

Then:

$$Nu_1 = \frac{2\dot{q}_1''d}{(t_1 - t_m)k} \quad ; \quad Nu_2 = \frac{2\dot{q}_2''d}{(t_2 - t_m)k}$$

9-2

$$Nu_i = \frac{8(\eta - 1)(1 - \eta)^2}{\eta\left[3 + 4\ln\eta - 4\eta^2 + \eta^4\right]}$$

where $\eta = r_i/r_o$

For $\eta = 0.6$ $\quad Nu_i = 6.24$

9-3

$$t_m = 10 + 45.83x; \quad t_w = t_m + \dot{q}_o''D/(kNu)$$

$$\dot{q}_o'' = 6410 \text{ W/m}^2$$

Nu as a function of **x** is obtained from Table 9-13. The highest fluid temperature is 129 °C.

9-4

Nu = 4.00 for both sides. Temperature profile is linear. If viscosity is a function of temperature the velocity profile will be affected, but this has no influence on temperature profile or Nusselt number.

9-5

Total heat flux from inner surface is approximately 2550 W/m^2. Outer tube surface temperature is approximately 278 ^oC. 55 percent of the heat from the fluid is transferred directly from the inner surface, and 45 percent indirectly from the outer surface.

9-6

Mean fluid temperature = **100 + 75(x/L) + 61(1-cos(πx/L))** ^oC

Surface temperature = **110 + 75(x/L) + 27sin(πx/L) + 61(1-cos(πx/L))**

Re = 1650, **Nu** = 8.24, **h** = 57 W/(m^{2}K)

9-7

Let the centerline temperature be t_c. Then,

$$t = t_c + \frac{4V^2 Pr}{c}\left[\left(\frac{r}{r_o}\right)^2 - \frac{1}{2}\left(\frac{r}{r_o}\right)^4\right]$$

Re = 1600, so flow is laminar.

Axial temperature gradient: $$\frac{dt}{dx} = \frac{8V\alpha Pr}{c r_o^2}$$

9-8

(The clearance should be 0.25 mm rather than 0.025 mm.)

$$t = t_h + \frac{\mu}{k} u_j^2\left[\left(\frac{y}{d}\right) - \frac{1}{2}\left(\frac{y}{d}\right)^2\right]$$

where t_h is the bearing surface temperature
u_j is the journal surface velocity
d is the clearance

$$\dot{q}_0'' = \frac{\mu u_j^2}{d} \approx 15500 \text{ W/m}^2$$

$$P = \frac{\mu u_j^2 \pi D L}{d}$$

Journal surface temperature $\approx 97^\circ C$

9-9

$$\dot{q}_o''(x^+) = \frac{2k}{r_o}\sum_n G_n e^{-\lambda_n^2 x^+}\left(a + b e^{x_i^+}\right), \qquad x^+ > x_i^+$$

$$\theta_m(x^+) = 1 - \frac{8}{a+b}\sum_n \left(a + b e^{\lambda_n^2 x_i^+}\right)\left(1 - e^{-\lambda_n^2 x^+}\right)\frac{G_n}{\lambda_n^2}, \qquad x^+ > x_i^+$$

9-10

$$\frac{t_o(x^+) - t_e}{(r_o/k)\dot{q}_o''} = 4x^+ + \sum_m \frac{1 - e^{-\gamma_m^2 x^+}}{\gamma_m^4 A_m}, \qquad 0 < x^+ < 0.1$$

$$\frac{t_o(x^+) - t_e}{(r_o/k)\dot{q}_o''} = 0.4 + \sum_m \frac{e^{-\gamma_m^2(x^+ - 0.1)} - e^{-\gamma_m^2 x^+}}{\gamma_m^4 A_m}, \qquad x^+ > 0.1$$

where $\dot{q}_o''$ is the constant heat flux for $0 < x^+ < 0.1$

$$\frac{t_m(x^+ - t_e)}{(r_o/k)\dot{q}_o''} = 4x^+, \qquad 0 < x^+ < 0.1$$

9-11

This problem involves the same procedure as 9-10 except that the procedure is repeated at x^+ intervals of 0.02. A local Nusselt number is readily evaluated from:

$$Nu = \frac{2(r_o/k)\dot{q}_o''}{(t_o - t_m)}$$

After a few steps **Nu** will repeat with little change in each heated segment.

9-12

x^+	Nu_x	Nu_m
0.01	4.55	9.03
0.02	4.10	6.63
0.05	3.42	4.88
0.10	3.07	4.03
0.20	2.98	3.53
1.00	2.98	3.53
5.00	2.98	3.00

Nu_x comes within 2% of its asymptotic value at $x^+ = 0.118$.

9-13

This problem involves substitution into Eq. (9-55) and evaluation of the infinite series terms (an exercise in computer programming). A Nusselt number is defined as:

$$\mathrm{Nu} = \frac{2(r_0/k)q_0''}{(t_0 - t_m)}$$

and t_m is obtained from Eq.(9-54). The Nusselt number goes to **0.0** at **x = L**, and the student should explain this behavior by sketching the presumed temperature profile at **x = L**.

9-14

$$t_m = t_e + \frac{2}{\rho c V r_0}\left[ax + \frac{bL}{\pi}\left(1 - \cos\frac{\pi x}{L}\right)\right]$$

$$t_0 = t_e + \frac{4 r_0}{k}\left[\frac{ax}{c} + \frac{bL}{c\pi}\left(1 - \cos\frac{\pi x}{L}\right)\right] +$$

$$\frac{r_0 a}{k}\sum_m \frac{\left(1 - e^{-\gamma_m^2 x/c}\right)}{A_m \gamma_m^4} + \frac{r_0 b}{k}\sin\frac{\pi x}{L}\sum_m \frac{1}{A_m\left(\gamma_m^4 + \frac{c^2\pi^2}{L^2}\right)} -$$

$$\frac{r_0 b}{k}\frac{c\pi}{L}\cos\frac{\pi x}{L}\sum_m \frac{1}{\gamma_m^2\left(\gamma_m^4 + \frac{c^2\pi^2}{L^2}\right)} + \frac{r_0 b}{k}\frac{c\pi}{L}\sum_m \frac{e^{-\gamma_m^2 x/c}}{\gamma_m^2 A_m\left(\gamma_m^4 + c^2\pi^2/L^2\right)}$$

where $c = r_0 \mathrm{RePr}$

9-15

Eqs.(9-56) and (9-57) provide the basis for the procedure desired here. The first integral in (9-56) and the second integral in (9-57) would be used. The integrations must be carried out numerically, using the basic data from Table 9-11. Note that at each station x^+, integration must be carried out from $\xi = 0$ to $\xi = x^+$. Note that, for example, if $x^+ = 1.0$ and $\xi = 0.8$, then $\mathrm{Nu}_{ii}(x^+ - \xi) = \mathrm{Nu}_{ii}(0.2) = 6.19$ for $r^+ = 0.5$.

9-16

This problem is designed as an application of Eq. (9-23). The source of plane radiation will produce a heat flux into the tube that varies sinusoidally around the periphery. The re-radiation will produce a heat flux from the tube that subtracts slightly from the inflow on one side, and results in net radiation <u>from</u> the tube on the other. The complete heat flux distribution is then approximated by:

$$\dot{q}_0'' = \dot{q}_a''(1 - b\cos\phi)$$

The tube-wall temperature is evaluated from:

$$t_0(\phi) = t_m + \frac{2\dot{q}_0''(\phi) r_0/k}{\mathrm{Nu}(\phi)}$$

9-17

This problem is solved by substituting the proposed expression for t_0 into Eq.(9-44) and carrying out the indicated integration (the summation term in (9-44) is not used.) The function θ is supplied by Eq. (9-36). The wall heat flux is evaluated as in Eq.(9-37), and the Nusselt number as in Eq.(9-33). For large values of x^+ it will be found that certain infinite series terms drop out leading to a result that is independent of x^+. The results are

$$Nu(x^+) = \frac{4\left[\sum_n \frac{G_n e^{bx^+}}{(\lambda_n^2 + b)} - \sum_n \frac{G_n e^{-\lambda_n^2 x^+}}{(\lambda_n^2 + b)}\right]}{\frac{1}{b}(e^{bx^+} - 1) - 8\sum_n \left[\frac{G_n}{b(\lambda_n^2 + b)}(e^{bx^+} - 1) + \frac{G_n}{\lambda_n^2(\lambda_n^2 + b)}(e^{-\lambda_n^2 x^+} - 1)\right]}$$

$$Nu(\infty) = \frac{4b\sum_n \frac{G_n}{\lambda_n^2 + b}}{1 - 8\sum_n \frac{G_n}{\lambda_n^2 + b}}$$

9-18

$$Nu = \frac{\frac{V r_0^2}{\alpha}\left(\frac{dt_m}{dx}\right) - \frac{8 V^2 \mu}{k}}{\frac{11}{48}\frac{V r_0^2}{\alpha}\left(\frac{dt_m}{dx}\right) - \frac{5}{6}\frac{V^2 \mu}{k}}$$

where $$\left(\frac{dt_m}{dx}\right) = \frac{2 q_0''}{r_0 \rho V c} + \frac{8 V \mu}{\rho c r_0^2}$$

9-19

$$Nu = \frac{1}{\frac{11}{48} + \frac{1}{16}\left(\frac{S}{\dot{q}_0''}\right) r_0} = \frac{48}{11}\left[\frac{1}{1 + \frac{3}{11}\left(\frac{S r_0}{q_0''}\right)}\right]$$

$$t = t_0 + \frac{V}{\alpha}\left(\frac{dt_m}{dx}\right)\left[\frac{r^2}{2} - \frac{r^4}{8r_0^2} - \frac{3}{8}r_0^2\right] + \frac{S}{4k}[r_0^2 - r^2]$$

$$\left(\frac{dt_m}{dx}\right) = \frac{2\dot{q}_0''}{\rho V c r_0} + \frac{S}{\rho V c}$$

9-20

The highest temperature occurs at the wall. The energy equation may be put in the form:

$$\frac{\partial^2 t}{\partial r^2} + \frac{1}{r}\frac{\partial t}{\partial r} = \frac{S}{k}\left[2\left(1 - \frac{r^2}{r_0^2}\right) - 1\right]$$

By L'Hospital's rule:

$$\lim_{r \to 0} \frac{1}{r}\frac{\partial t}{\partial r} = \left.\frac{\partial^2 t}{\partial r^2}\right|_0$$

Thus at **r = 0,**

$$\left.\frac{\partial^2 t}{\partial r^2}\right|_0 = \frac{S}{2k} > 0$$

9-21

This problem is an unheated starting length problem, created by starting the calculations with a fully-developed velocity profile and a flat temperature profile. Where the flow begins, the wall temperature will be some constant value different from the inlet temperature. Let the Reynolds number be 1000, and let the pipe radius be 4 cm. From the information in Table 9-4 $x^+ > 0.3$ will be the completely thermally fully-developed length for the heat transfer. Based on the value 0.15, the resulting pipe length, in meters, will be **x** = 12.0***Pr**. Note, this is excessive for the large Prandtl numbers, so use a very large numerical stepsize.

The Nusselt number distributions for the 3 Prandtl number cases plot on top of each other throughout the entry region, when plotted versus , and they asymptote out to 3.16 for this constant wall temperature problem by about $x^+ = 0.15$. Thus, there is negligible effect of Prandtl number on the unheated starting length, when considered in the x^+ coordinate.

Plotting the non-dimensional temperature profiles, $(t_o - t)/(t_o - t_m)$, shows the evolution of the flat profile at start of heating/cooling, and the evolution into a shape similar to that depicted in Figure 9-4 for the constant wall temperature. By $x^+ = 0.05$, the profile is near the non-changing thermally fully-developed shape, and it is almost unchanging by 0.15, where the criterion from Eq. (9-7) is satisfied.

9-22

This problem is an unheated starting length problem, created by starting the calculations with a fully-developed velocity profile and a flat temperature profile. Where the flow begins, the wall heat flux will be some constant value. Use the same geometry as described in Problem 9-21, and consider a wall heat flux of 10 W/m^2. Discussion of the results pretty much follows 9-21, except that the final shape of the non-dimensional temperature profile will be different, as depicted in Figure 9-4.

9-23

This problem is an unheated starting length problem that is essentially a combination of the Problems 9-21 and 9-22, but applied to flow between parallel plates. Let the Reynolds number be 1000, and let the channel width be 2 cm. From the information in Tables 9-9 and 9-10 $x^+ > 0.3$ will be the completely thermally fully-developed length for the heat transfer.

The discussions in Problems 9-21 and 9-22 pertain to this problem as well, although the thermally fully-developed profile shapes for the two different thermal boundary conditions are not quite as pronounced. The approach to being thermally fully-developed is somewhat faster than for the pipe.

9-24

The geometry for this problem is specified in Problem 9-3. Be sure to use constant properties in the calculation, regardless of whether it may be justified. This is a combined entry problem in that the initial profiles are flat. From an energy balance (see the solution to Problem 9-3) the wall heat flux will be 6410 W/m^2 to heat the fluid from 10 to 65 deg C. From the mass flow rate and geometry, the Reynolds number is 411.

The Nusselt number distribution in the tube will be nearly infinite, decreasing to nearly constant by $x^+ = 0.05$ (the tube exit for this Prandtl number). Note the friction coefficient will become constant very near the entrance, making this basically an unheated starting length problem with a constant heat flux boundary condition. Compare your results with Table 9-6.

9-25

The geometry for this problem is specified in Problem 9-6. Be sure to use constant properties evaluated at 250 deg C. Use a Reynolds number, based on this temperature, of 1625. This is a combined entry problem in that the initial profiles are flat. The thermal boundary condition is the variable heat flux of Problem 9-6.

Results of the calculations show the heat transfer coefficient becomes essentially constant by the first 10 percent of the tube, whose overall length is $x^+ = 0.35$. The results of the variable heat flux calculation yield a wall and mean temperature distribution that is similar to that in solution 9-6. The exit mean temperature is 634 deg K (361 deg C). The wall temperature peaks just before the exit and begins to fall. This shows the significant lag in the wall temperature compared to the drop in wall heat flux, characteristic of laminar flows.

9-26

This problem begins as a combined entry problem with flat velocity and temperature profiles. Let the Reynolds number be 1000. For air at room temperature, let the pipe radius be 4 cm, and let the length be 5.7 m, with the wall temperature constant at 50 deg C for the first 2.85 m, then increased to 60 deg C, and then linearly decreased back to 50 deg C over the last half of the pipe. In the discussion which follows, $\mathbf{x^+}$ is measured from the start of the pipe, with the step wall-temperature change at $\mathbf{x^+} = 0.2$, and the end of the pipe at $\mathbf{x^+} = 0.4$.

The combined entry Nusselt number can be compared with Table 9-12. The asymptotic behavior of **Nu** will abruptly change to positive large at the start of the step-change in wall temperature, then approach negative large at $\mathbf{x^+} \approx 0.32$, re-emerging as positive large, and decreasing. This follows the trend shown in Figure 9-14. The wall heat flux begins positive large at $\mathbf{x^+} = 0$ and decreases, until it once again becomes positive large at $\mathbf{x^+} = 0.2$, followed by a monotonic decrease through zero at about $\mathbf{x^+} = 0.3$ and becoming negative. This crossover to negative comes prior to the location $\mathbf{x^+} = 0.33$ where the decreasing wall temperature equals the bulk mean temperature. This is the location of the indefinite Nusselt number (changing from negative infinite to positive infinite). To further understand this behavior, the non-dimensional temperature profiles need to be examined in the vicinity of $\mathbf{x^+} = 0.33$. Note that the maximum temperature moves from the centerline and becomes significantly inflected at $\mathbf{x^+} = 0.37$. This reflects the slow adjustment of the flow to the changing thermal boundary condition.

Chapter 10

10-1

Follow the procedure on Page 165. Integrate 3 times and apply boundary conditions.

Pr <<1

$$\tau = \frac{\int_0^\eta \exp\left(-\frac{Pr}{4}\eta^2\right)d\eta}{\int_0^\infty \exp\left(-\frac{Pr}{4}\eta^2\right)d\eta} = erf\left(\frac{\sqrt{Pr}}{2}\eta\right)$$

The Nusselt number is

$$Nu_x = \tau'(0)Re_x^{1/2}$$

$$Nu_x = \frac{1}{\sqrt{\pi}}Pr^{1/2}Re_x^{1/2}$$

Pr<<1

$$\tau = \frac{\int_0^\eta \exp\left(-\frac{B}{3}\eta^3\right)d\eta}{\int_0^\infty \exp\left(-\frac{B}{3}\eta^3\right)d\eta}; B = \frac{0.3321}{4}Pr$$

To evaluate the integrals, transform the equations using $\xi = \frac{B}{3}\eta^3$ to obtain Gamma Functions. Then:

$$Nu_x = 0.339\,Pr^{1/3}Re_x^{1/2}$$

10-2

Eq. (10-9) must be evaluated from **Pr** = 0.01. The $\zeta(\eta)$ are given in Table 8-1. Numerical integration of (10-7) for C_1 yields $C_1 = \tau'(0) = 0.0519$. This is about 9% lower than the value in (10-11) for **Pr** = 0.01, primarily because **u** = **u**(η) in the solution.

10-3

The applicable d.e. is that for $u_\infty = Cx^m$, page 166. Then for **Pr << 1**, $\zeta' = 1$. The resulting d.e. is,

$$\tau'' + \left(\frac{m+1}{2}\right)Pr\,\eta\,\tau' = 0$$

Integrate and form a Nusselt number:

$$Nu_x = \frac{Re_x^{1/2}}{\int_0^\infty \exp\left(-\frac{Pr}{2}\eta^2\right)d\eta} = \frac{Re_x^{1/2}}{\sqrt{\frac{\pi}{2Pr}}}$$

$$Nu_x = 0.798\,Pr^{1/2}\,Re_x^{1/2}$$

For the stagnation solution in terms of **D**, using Eq. (10-20),

$$Nu_D = 1.596\,Pr^{1/2}\,Re_D^{1/2}$$

10-4

The Nu_x equation, similar to (10-9), but for $u_\infty = Cx^m$ is:

$$Nu_x = \frac{Re_x^{1/2}}{\int_0^\infty \exp\left[-\left(\frac{m+1}{2}\right)Pr\int_0^\eta \varsigma d\eta\right]d\eta}$$

Numerically integrate, to yield:

$$Nu_x = 0.0759\,Re_x^{1/2}$$

10-5

The d.e. is:

$$\tau'' + \frac{Pr}{2}(m+1)\varsigma\tau' - Pr\,m\,\varsigma'\,\tau = 0$$

The integration of this equation is not straight-forward because it is a boundary value problem. A solution can be found in NACA TN3151 and TN3588 by Livingood and Donoughe. Other solutions are referenced in Schlichting, page 288. For **Pr = 0.7** the result is:

$$Nu_x = 0.4803\,Re_x^{1/2}$$

10-6

The solution for $\dot{q}_0''$ is Eq.(10-33). Let **a = 0, b = 1**. To evaluate the integral, transform the equation letting $w = 1-(\xi/x)^{3/4}$.

$$\int_0^x \left[1-\left(\frac{\xi}{x}\right)^{3/4}\right]^{-1/3} d\xi = -\left(\frac{4}{3}x\right)\int_1^0 w^{-1/3}(1-w)^{1/3}dw$$

$$= \frac{4}{3}x\frac{\Gamma(2/3)\Gamma(4/3)}{\Gamma(2)}$$

$$= 1.6123x$$

The final result is:

$$Nu_x = 0.534\,Pr^{1/3}\,Re_x^{1/2}$$

10-7

The conduction thickness is $\Delta_4 = k/h$. From Problem 10-3 and Eq. (10-20),

$$\Delta_4 = \frac{k}{h} = \frac{x}{Nu_x} = \sqrt{\frac{\pi \nu x}{2 Pr u_\infty}} = \sqrt{\frac{\pi}{8}\frac{\nu}{Pr}\frac{D}{V}}$$

Then Δ_4 = **1.049 mm**. The shear thickness, from Eq.(8-36):

$$\delta_4 = \frac{u_\infty}{(\partial u/\partial y)_0} = \frac{1}{\varsigma''(0)}\sqrt{\frac{\nu x}{u_\infty}} = \frac{1}{\varsigma''(0)}\sqrt{\frac{\nu D}{4V}}$$

from which δ_4 = **0.0584** mm.

10-8

The first task is to perform an energy balance on a unit of surface area in the stagnation region. Let t_0 be the wall, t_i be the injection or internal coolant temperature.

Case I: Blowing, no radiation effect:

$$\dot{m}_0'' c(t_0 - t_i) - h(t_\infty - t_0) = 0$$

Case II: Blowing plus surface radiation:

$$\dot{m}_0'' c(t_0 - t_i) - \sigma(T_\infty^4 - T_0^4) - h(t_\infty - t_0) = 0$$

Case III: Internal convective cooling:

$$\dot{m}_0'' c(t_0 - t_i) - h(t_\infty - t_0) = 0$$

For the stagnation point Nusselt number, **m = 1**, use Table 10-4. Convert Nu_x, Re_x to Nu_D, Re_D, using (10-20). For Case III, $v_0 = 0$. Case II requires iteration. For Case III, $\dot{m}_0''$ varies while **h** is a constant. Carry out calculations for $0 \le \dot{m}_0'' \le 0.13$ kg/(cm2).

Results:

$\dot{m}_0''$	h	t_0(I)	t_0(II)	t_0(III)
0	94	540	540	540
.0644	55	265	382	328
0.129	28	125	250	240

10-9

The heat flux is given by Eq. (10-31) with $dt_0/d\xi = 0$ and $h(\xi,x)$ given by (10-32). Thus:

for **0.00 < x < 0.01 m**

$$\dot{q}_0'' = \frac{0.332k}{x} Pr^{1/3} Re_x^{1/2} \Delta t_{0,1}$$

for **0.01 < x < 0.02 m**

$$= \frac{0.332k}{x} Pr^{1/3} Re_x^{1/2} \left[\Delta t_{0,1} + \left[1 - \left(\frac{0.01}{x}\right)^{3/4}\right]^{-1/3} \Delta t_{0,2} \right]$$

for $0.02 < x < 0.03$ m

$$= \frac{0.332k}{x} Pr^{1/3} Re_x^{1/2} \left[\Delta t_{0,1} + \left[1 - \left(\frac{0.01}{x}\right)^{3/4}\right]^{-1/3} \Delta t_{0,2} + \left[1 - \left(\frac{0.02}{x}\right)^{3/4}\right]^{-1/3} \Delta t_{0,3} \right]$$

where $\Delta t_{0,1} = -50\ ^oC$
$\Delta t_{0,2} = +40\ ^oC$
$\Delta t_{0,3} = -40\ ^oC$

Results:

x/L	$\dot{q}_0''$(W/m^2)	h(W/(m^{2}K)
.0833	-1850	33
.167	-1310	24
.25	-1070	19
.333	-930	17
.35	1580	-142
.417	410	-37
.5	190	-17
.583	100	-9
.667	70	-7
.683	-1920	35
.75	-1110	20
.883	-870	16
.917	-760	14
1.0	-690	13

10-10

The heat flux is given by Eq.(10-31) with $h(\xi_1,x)$ given by (10-32). Now $\Delta t_0 = (t_0 - t_\infty)\big|_{\xi=0} = -50°C$ and $(t_0 - t_\infty) = 40\sin\frac{\pi\xi}{L}$ for $0 < x < L$, where $L = 0.03$ m.

$$\dot{q}_0'' = \frac{0.332}{x} k\, Pr^{1/3} Re_x^{1/2} \left\{ \int_0^x \left[1 - \left(\frac{\xi}{x}\right)^{3/4}\right]^{-1/3} \frac{40\pi}{L} \cos\frac{\pi\xi}{L} d\xi - 50 \right\}$$

The integral can be evaluated by transforming using $u = (\xi/x)^{3/4}$ and expanding the cosine in a power series.

$$I = \frac{4}{3}x\int_0^1 u^{1/3}(1-u)^{-1/3}\cos\frac{\pi x}{L}u^{4/3}\,du$$

$$= \frac{4}{3}x\left\{\sum_{k=0}^{\infty}\frac{(-1)^k\left(\frac{\pi x}{L}\right)^{2k}}{(2k)!}\beta\left[\frac{4}{3}(1-2k),2/3\right]\right\}$$

The solution is then programmed and the first ten terms of the series is sufficient:

x/L	$\dot{q}''_0$ (W/m²)	h(W/m²K)
.1	-1180	28
.2	-360	12
.3	-2	12
.4	90	-7
.5	30	-3
.6	-130	10
.7	-380	19
.8	-670	23
.9	-960	24
1.0	-1220	22

10-11

Eq.(10-52) is applicable. Let $\theta = x/R$ and $u_\infty = 2V\sin(x/R)$ to transform from **x** to θ in (10-52). Then:

$$Nu_D = 0.836\,Pr\,Re_{D,v}^{1/2}\frac{(\sin\theta)^{1.435}}{\left(\int_0^\theta(\sin\theta)^{1.87}d\theta\right)^{1/2}}$$

For $\theta \to 0$, $\sin\theta \to \theta$ and $Nu_D \to 0.99\,Re_{D,v}^{1/2}$

θ	$Nu_D/Re_{D,v}^{1/2}$
0	.99
.05p	.98
.1p	.96
.2p	.92
.3p	.86
.4p	.72
.5p	.65

10-12

Similar results as 10-11, except that the leading coefficient changes:

$$Nu_D = 0.724\,Pr\,Re_{D,v}^{1/2}\frac{(\sin\theta)^{1.435}}{\left[\int_0^{\theta}(\sin\theta)^{1.87}d\theta\right]^{1/2}}$$

10-13

Eq. (10-41) is applicable. Let x_3 = **15** cm and $\dot{q}_0''$ be the required heat flux. Then,

$$t_0(x_3) - t_\infty = \frac{0.623}{k}Pr^{1/3}Re_x^{-1/2}\left[\int_0^{x_1}(0)d\xi + \int_{x_1}^{x_2}\left[1-\left(\frac{\xi}{x}\right)^{3/4}\right]^{-2/3}\dot{q}_0''d\xi + \int_{x_2}^{x_3}(0)d\xi\right]$$

Transform the integral into an Incomplete Beta Function using $u = 1-(\xi/x)^{3/4}$. Then the integral becomes:

$$I = \dot{q}_0''\left(-\tfrac{4}{3}x_3\right)\int_{u/\xi=x_1}^{u/\xi=x_2}u^{2/3}(1-u)^{1/3}du$$

$$= \dot{q}_0''\left(\tfrac{4}{3}x\right)\left[\int_{u_2}^{0}u^{m-1}(1-u)^{n-1}du + \int_0^{u_1}u^{m-1}(1-u)^{n-1}du\right]$$

where **m=1/3, n=4/3**. For $t_0(x_3)$ = **0°C**, $\dot{q}_0'' \cong$ **1600** W/m2. Eq.(10-41) is then used to find $t_0(x)$ for **x > 7.5** cm.

10-14

For **x > 15** cm, Eq.(10-41) applies. For $0 < x \le 15$**cm**,

$$\dot{q}''(\xi) = 0.292k(t_0 - t_\infty)\left(\frac{u_\infty\xi}{\nu}\right)^{-1/2}$$

where $(t_0 - t_\infty) = 210$°C. For $15 < x \le 60$ cm, $\dot{q}_0''$ = **0.** The integral can then be transformed using $u = 1-(\xi/x)^{3/4}$. Then:

$$(t_0 - t_\infty) = \frac{(0.623)(0.292)(-210)}{Pr^{1/3}}\frac{4}{3}\left[\int_0^1 u^{-2/3}(1-u)^{-1/3}du - \int_0^{1-\left(\frac{1}{2x}\right)^{3/4}} u^{-2/3}(1-u)^{-1/3}du\right]$$

The first integral is $\beta_1(1/3,2/3)$ and the second integral is $\beta_r(1/3,2/3)$. Then:

x (cm)	t_0(°C)
15.71	140
16.5	158
17.5	183
20.5	194
30.1	223
51.8	246
130	268
∞	300

10-15

Assume the boundary layer begins at the start of the convergence. The applicable equation is (10-54). The solution procedure is similar to Prob. 8-5. approximate $\mathbf{u}_\infty$ as linear with **x.** The solution must be numerically integrated, resulting in:

θ	**h** (W/m²K)
15	790
30	990
45	1730
60	2210
75	2470
80	(**Nu_D** = 27.5, **max**)
90	2800

10-16

Set up the problem for laminar flow over a flat plate with free stream velocity of 15 m/s, free stream temperature of 300 K, and wall temperature be 295 K. Assume constant properties and start the calculations at an x-Reynolds number of about 1000. For **Pr** = 0.7 and 1.0, let the fluid be air at 1 atm. For this case a plate length of 20 cm is sufficient. For **Pr** = 5.0, let the fluid be water, with a plate length of 1 cm. Evaluate the initial velocity profile by converting the Table 8-1 values from their similarity form to physical variables. The thermal profile will have to come from another literature source, or you can use an approximate profile, such as Eq. (10-25).

The computations should go out to a $\mathbf{Re_x}$ of about 2 x 10^5, and the $\mathbf{Re_{\delta 2}}$ should range from an initial value of about 20 out to about 290, independent of the heat transfer for the constant property solution. For **Pr** = 0.7 and 1.0, the $\mathbf{Re_{\Delta 2}}$ starts at 26 and grows to 365 and 290 respectively. The latter confirms the Reynolds analogy idea from Eq. (10-3) where $\mathbf{c_f/2}$ = **St** and $\mathbf{Re_{\delta 2}}$ grows at the same rate as $\mathbf{Re_{\Delta 2}}$. For **Pr** = 5.0, the $\mathbf{Re_{\Delta 2}}$ grows to 100, indicating the thermal boundary layer resides inside the momentum boundary layer. For each **Pr**, log-log plot **St** versus $\mathbf{Re_x}$ to compare with Eq. (10-10) and versus $\mathbf{Re_{\Delta 2}}$ to compare with Eq. (10-16). From the enthalpy thickness distribution, create its **x**-derivative and compare the results to Eq. (6-24) for this constant wall temperature boundary condition.

10-17

Set up the problem for laminar flow over a flat plate with free stream velocity varying as $\mathbf{Cx^m}$ (m/s) and free stream temperature of 300 K. Set up the problem for laminar flow over a flat plate with free stream velocity varying as $\mathbf{Cx^m}$ (m/s), free stream temperature of 300 K, and wall temperature be 295 K. Assume constant properties.

For **m**=0.111, start the calculations at an **x**-Reynolds number of about 1000. For **Pr** = 0.7 and 1.0, let the fluid be air at 1 atm, and for 5.0, let the fluid be water. Stop the calculations at an **x**-Reynolds number of about 200,000. See the write-up for Problem 8-8 for details on the free stream velocity distribution and initial profiles. For this **m** value, C = 592 yields a starting location of 0.0001 m and an ending location of 0.012 m. For **m**=1.0, the same **C** yields a starting **x**-Reynolds number of about 100 at a **x** value of 0.002 m and an ending location of 0.12 m yields an ending **x**-Reynolds number of about 300,000.

The computations have limited comparisons, but you should observe the sqrt(**x**) variation of the variables, and the friction coefficient can be compared to Eq. (8-25) along with Table 8-2. The Nusselt numbers can be compared to Table 10-2.

The computations will have limited comparisons, but you should observe the sqrt(**x**) variation of the variables, and the friction coefficient can be compared to Eq. (8-25) along with Table 8-2. For **m**=0.111, the shape factor will be 2.47 and for **m**=1.0, the value will be 2.21. The Stanton number (or Nusselt number) can be computed using the energy integral Eq. 6-18. Similarity can be confirmed by plotting temperature profiles at different x-locations, as a function of Prandtl number. Unlike the **m**=0 solution, the growth of $\mathbf{Re_{\delta 2}}$ is not at the same rate as $\mathbf{Re_{\Delta 2}}$ for **Pr**=1.0. Confirm that for **m**=1.0, the heat transfer coefficient does not change with **x**.

10-18

Set up the problem for laminar flow over a flat plate with free stream conditions, properties, and thermal boundary condition as specified by Problem 10-9. Evaluate the properties at the free stream temperature, although this could also be considered a variable property problem. It is not crucial where you start the calculations, as long as it is within the first one-third of the plate. This is a laminar flow calculation, use Blasius velocity and temperature profiles.

The results of the calculations can be easily compared to the solution to 10-9. Although you may find your heat flux numbers and heat transfer coefficients to be somewhat different, they should agree to within about 10-20 percent. Part of the disagreement may be due to the assumed properties. If you wish, you may adjust your properties to reproduce the initial heat flux table entry shown in the solution to 10-9, and this will improve your comparison, but in any event, the comparison between theory and numerical results are quite good, including near the step change in wall temperature, which causes the heat flux and heat transfer coefficients to change sign.

10-19

Set up the problem for laminar flow over a flat plate with free stream conditions and properties as specified by Problem 10-9. For the thermal boundary conditions, as specified by Problem 10-10, let $\mathbf{t_o} = \mathbf{a} + \mathbf{b} \sin (\pi\ \mathbf{x/L})$ where $\mathbf{a}$ = 40 deg C (313 K), $\mathbf{b}$ = 40 (deg C or K), and $\mathbf{L}$ = 0.3 m. Evaluate the properties at the free stream temperature, although this could also be considered a variable property problem. For this problem, where the wall temperature varies, it is important that you start the calculations near the leading edge of the plate. This is a laminar flow calculation, use Blasius velocity and temperature profiles.

The results of the calculations can be easily compared to the solution to 10-10. As with Problem 10-18, you may find your heat flux numbers and heat transfer coefficients to be somewhat different, although they should agree to within about 10-20 percent. Part of the disagreement may be due to the assumed properties. If you wish, you may adjust your properties to reproduce the initial heat flux table entry shown in the solution to 10-10, and this will improve your comparison, but in any event, the comparison between theory and numerical results are quite good. To understand the phase lag between the heat flux changing sign and the heat transfer coefficient changing sign, you need to plot temperature profiles (the region around **x/L** = 0.4- 0.5).

10-20

Set up the problem for laminar flow over a flat plate with free stream conditions, properties, and thermal boundary condition as specified by Problem 10-13. Evaluate the properties at the free stream temperature. It is not crucial where you start the calculations, as long as it is within the first 5 cm of the plate. This is a laminar flow calculation, use Blasius velocity profile and a flat temperature profile (wall is adiabatic). You will, in essence, have to solve Problem 10-13 to know what heat flux to specify over the interval of **x** from 0.05 to 0.75 cm.

The results of the calculations can be compared to the solution to 10-13. Although you may find your wall temperature distribution to be somewhat different, it should agree to within about 10-20 percent. Again, part of the disagreement may be due to the assumed properties. Observe the rise in temperature and fall in heat flux over the heating portion of the plate, and the decay of temperature over the adiabatic portion. You may find your calculated heat flux gives a temperature at 15 cm that is not quite 0 deg C, but it should be within 1-2 degrees.

Chapter 11

11-1

The equation which must be numerically integrated is :

$$\frac{du^+}{dy^+} = \frac{2}{1+\sqrt{1+4\kappa^2 y^{+2} D^2}}$$

where $D = 1 - e^{-y^+/A^+}$

The following table lists some values of $u^+(y^+)$ which will be obtained using various value of A^+:

u^+ for :

y^+	$A^+ = 22$	25	27	Eq.(11-16)
59.59	14.41	15.01	15.40	14.97
79.32	15.13	15.75	16.14	15.67
95.97	15.60	16.22	16.62	16.14

11-2

For

$$y^+ \le y^+_{crit}, \quad u^+ = \frac{1}{v_0^+} e^{v_0^+ y^+} - \frac{1}{v_0^+}$$

For

$$y^+ > y^+_{crit}, \quad u^+ = \frac{v_0^+}{4\kappa^2}\left(\ln\frac{y^+}{y^+_{crit}}\right)^2 + \frac{1}{\kappa} e^{\frac{v_0^+ y^+_{crit}}{2}} \ln\frac{y^+}{y^+_{crit}} + \frac{1}{v_0^+} e^{v_0^+ y^+_{crit}} + \frac{1}{v_0^+}$$

For

$v_0^+ = 0.1773, \quad y^+_{crit} \approx 4.4$

$v_0^+ = -0.065, \quad y^+_{crit} \approx 35$

11-3

The equation which must be numerically integrated is:

$$\frac{du^+}{dy^+} = \frac{2(1+v_0^+ u^+)}{1+\sqrt{1+4\kappa^2 y^{+2} D^2 (1+v_0^+ u^+)^2}}$$

where $D = 1 - e^{-y^+/A^+}$

For

$v_0^+ = 0.1773, \quad A^+ \approx 10.2$

$v_0^+ = -0.065, \quad A^+ \approx 80$

Eq. (11-26) gives close to the same result for the first case, but gives 60 for the second. This is simply the result of fairly large experimental uncertainty for the experiments with strong suction.

11-4

In the turbulent region:

$$C_f/2 = 0.0287\left[\mathbf{Re_x}+\left(37\mathbf{Re_t^{0.625}}-\mathbf{Re_t}\right)\right]^{-0.2}$$

where $\mathbf{Re_t}$ is the transition Reynolds number.

The virtual origin of the turbulent boundary layer will be at:

$$x = \frac{\nu}{u_\infty}\left[\mathbf{Re_t} - 37\,\mathbf{Re_t^{0.625}}\right]$$

11-5

$$\delta_2 = \frac{0.036}{\rho R u_\infty^{3.29}}\left[\int_0^x \mu^{.25}\rho R^{1.25} u_\infty^{3.86}\,dx\right]^{0.8}$$

11-6

z, cm (axial)	x, cm	P, Pa	T, K	δ_2, m	$\mathbf{Re}_{\delta 2}$
0	0	2089500	2473	0	0
1	1.38	2087400	2471	6.77E-5	70.5
2	2.75	2085200	2468	7.84E-5	94.8
3	4.13	2079000	2466	7.81E-5	111
4	5.50	2072700	2463	7.29E-5	123
5	6.88	2062200	2457	6.54E-5	134
6	8.25	2051700	2450	5.68E-5	144
8	11.0	1932000	2401	3.87E-5	164
----transition---					
10	13.7	1627500	2232	5.13E-5	369
* 12	15.7	1022070	1854	5.16E-5	504
14	17.5	449400	1332	7.50E-5	745
16	19.8	189400	943	1.56E-4	1362
18	22.1	95970	718	2.90E-4	2200
22	26.7	36120	485	6.59E-4	3854

*throat

$\dot{m} = 5.28$ kg/s

11-7

z, cm (axial)	x, cm	P, Pa	T, K	δ_2, m	$Re_{\delta 2}$
0	0	2100000	1275	0	0
----transition----					
1.56	2.15	2090613	1273	3.13E-5	209
2.36	3.22	2088135	1273	5.00E-5	375
3	4.13	2085135	1272	5.87E-5	489
4	5.5	2078958	1271	6.26E-5	627
6	8.25	2053065	1267	5.99E-5	906
8	11	1964634	1251	4.60E-5	1153
10	13.7	1672041	1195	3.77E-5	1577
* 12	15.7	1109388	1063	3.65E-5	2014
14	17.5	531846	861	5.19E-5	2683
18	22.1	139230	587	3.01E-4	10075
22	26.7	58905	459	6.19E-4	14158
28	33.6	23898	355	1.47E-3	22418
34	40.5	12117	292	2.88E-3	32082

*throat $\dot{m} = 18.7$ kg/s

11-8

This problem is a direct application of Eq.(8-42) for the laminar parts, and Eq.(11-21) for the turbulent parts. (The **R** cancels out in both cases.) Problems 8-4 and 11-5 should be solved first to get equation forms applicable to varying free-stream density. The laminar stagnation region near the leading edge is probably best handled by using Eq.(10-20) to evaluate the local velocity tangential to the surface, but just outside of the boundary layer. (Note that $\mathbf{u_\infty = 0}$ at $\mathbf{x = 0}$.) The momentum thickness at the stagnation point will be finite, not zero. Because there is a laminar boundary layer preceding the turbulent boundary layer, Eq.(11-21) should be modified by using the value of δ_2 from the laminar analysis as the lower limit on δ_2 when integrating the equation preceding (11-21).

x, cm (distance from stag. point)	$Re_{\delta 2}$ Upper surface	$Re_{\delta 2}$ Lower surface
0.2	82	59
0.5	140	105
1.0	200	152
2.0	430	260
3.0	820	420
5.0	1900	450
6.0	2200	-

11-9

$k_s \approx 0.05$ cm; $Re_k = 124$, Fully Rough.

To fit the theory the origin for **y** should be shifted a distance of 0.018 cm into the wall. The data then fit Eq. (11-56) very well, although of course there is a distinct "wake" region, virtually identical to that for a smooth surface. $c_f/2$ corresponds closely to Eq. (11-57).

11-10

Set up the problem for a constant free stream velocity of 10 m/s and a free stream temperature of 300 K. Let the properties be air at 1 atm, constant, and let the plate length be 10 m. Start the calculations at about 1.23 m. This should provide a momentum thickness Reynolds number range from 2,000 to 10,000. The initial velocity profile can be constructed as outlined in the problem statement. Analysis of the problem can include development of both the velocity profiles and the friction coefficient.

For the velocity profiles, consider $Re_{\delta_2} = 4000$ and confirm the viscous sublayer, Eq. (11-15) which extends out to about y^+ of 3.5, and the law-of-the-wall, Eq. (11-16), an well as the general shape of the profile in u^+ versus ln y^+ coordinates, as depicted in Figure 11-3 and 11-5. Re-plot the velocity profile in defect coordinates to observe the outer 80 percent of the profile, as shown in Figure 11-11, and evaluate the Clauser shape factor. Plot the profile using a power-law form, similar to that leading up to Eq. (11-18) to determine what the power law is for this profile. Be sure to use a boundary layer thickness that is reasonable, such as the location where $u/u_\infty = 0.99$. Evaluate the profile to determine δ_1 from Eq. (6-5) and δ_2 from Eq. (6-6), and evaluate the shape factor **H** (about 1.37 at this Reynolds number). Pay particular attention of the uncertainty in evaluation of the displacement and momentum thickness. If possible, plot the non-dimensional shear stress profile at $Re_{\delta_2} = 4000$ to confirm the Couette-flow part of the profile, Eq. (11-13).

Carry out the profile analyses at x-locations corresponding to $Re_{\delta_2} = 6000$, 8000, and 10000 to confirm both the inner-region similarity (u^+ versus ln y^+) and outer-region similarity (defect coordinates), and that **G** does not change as the boundary layer evolves. Observe how the power law exponent changes over the Re_{δ_2} range.

Plot the ln $c_f/2$ distribution versus ln Re_{δ_2} to confirm Eq. (11-20) and versus ln Re_x to confirm Eqs. (11-23) and (11-24). You can also determine $c_f/2$ by use of the momentum integral equation, (6-9). Be careful of accuracy in creating the momentum thickness derivative and its uncertainty.

If possible, modify the turbulence model (A^+ and κ) to see what effect it has on the friction results.

Chapter 12

12-1

$V/u_c = 0.875$

$c_f/2 = 0.041\ Re^{-0.25}$

where $Re = 2hV/\nu$
h is the plate spacing.

12-2

A numerical calculation using Eq. (12-6), but using the Van Driest mixing-length for $y^+ < 100$, $A^+ = 26$, and $\kappa = 0.40$, yields the following:

$Re = 29{,}100$, $c_f = 0.00302$
$Re = 102{,}307$, $c_f = 0.00220$

12-3

A solution can be obtained based on a slight modification of Eq.(11-21) to account for the difference between Eqs. (11-17) and (12-7). If the integral is evaluated numerically the following is a sample of the results:

$Re = 25000$ ($Re = 2dV/\nu$, d = plate spacing)

x/2d	$c_f/2$	V/u_c
1	.00409	.976
2	.00370	.959
3	.00352	.944
4	.00342	.931
5	.00337	.919
6	.00333	.908
7	.00331	.898
8	.00329	.888
9	.00329	.879
(entry length) 9.4	.00329	.875

Total pressure drop, $\dfrac{\Delta P}{(\rho V^2)} = 0.155$

12-4

The velocity profile is Eq. (11-56). This is substituted into Eq. (7-6). After integration is carried out to obtain a relation between the mean and center-line velocities, Eq. (12-17) is obtained from Eq. (11-56) when it is applied at the center-line.

12-5

Using Eq. (12-6) but employing the Van Driest mixing-length for $y^+ < 100$, $A^+ = 26$, $\kappa = 0.40$, the following is obtained:

Re	V/u_c
10103	0.789
29100	0.814
102307	0.837
311152	0.853
1013370	0.866

12-6

$$y_\ell^+ = 5; \left(\frac{y}{D}\right)_\ell \frac{DV\sqrt{C_f/2}}{\nu} = 5$$

$$\left(\frac{y}{D}\right)_\ell = \frac{5}{\sqrt{C_f/2}}\frac{1}{Re}$$

$$C_f/2 = 0.023 Re^{-0.2}$$

$$\therefore \quad \left(\frac{y}{D}\right)_\ell = 33 Re^{-0.8}$$

It will be found that the line on Fig. 12-2 labeled "limit of fully rough region" is a similar function of (k_s/D) [with a larger coefficient].

12-7

The necessary elevation difference is **1.21** m. The rough tube flow rate will be **0.45** kg/s, and the smooth tube rate **0.55** kg/s.

12-8

Set up the problem first for a value of **Re** = 30,000. Choose properties for any fluid; since this is a constant-property problem, the choice will not matter. The integration should be carried out from x = 0 to **x** > hydrodynamic entry length, so let $L/D_h \approx 75$ to assure hydrodynamically fully-developed flow. If you choose air at room temperature, then a pipe diameter 20 cm and a length of 15 m should be sufficient. This large L/D_h is because the centerline velocity will require much longer to develop than the wall friction. For the **Re** = 150,000 calculation, use the same radius. We suggest you use a turbulence model that is a composite of a mixing-length for the near-wall flow and an eddy viscosity model for the outer part of the flow. The procedure suggested in the textbook works quite well.

There are several features of the flow that can be compared to the chapter. Plotting the ratio u_{cl}/V shows a continual acceleration of the centerline velocity as the velocity profile changes shape. Note the change is significantly less than for laminar flow. It is worthwhile to plot a set of normalized velocity profiles, **u/V**, over $0 \le r/r_o \le 1$ to show how the initially flat profile changes. The centerline ratio will approach about 1.32 for the low **Re** and 1.26 for the higher **Re**. Carrying out the analysis leading to Eq. (12-10) using your computational results will allow you to evaluate the "computational experimental" exponent for the power-law form of the profile. Plotting the velocity profile in wall coordinates will permit comparison of the log-region part of the profile with Eqs. (12-4) and (12-6).

One very obvious feature is the rapid development of $c_f/2$ to its fully-developed value (about 15-20 hydraulic diameters for both **Re**). There exists several choices of fully-developed friction coefficient equations to evaluate the approach. Eq. (12-14) is quite accurate, although the comparison is quite sensitive to the details of the mixing-length/eddy viscosity turbulence model. When experimental uncertainty is considered, perhaps the choice of equation is not too important. If possible, evaluate the eddy diffusivity at the fully-developed state, and compare to Figure 12-1 and Eq. (12-5) for both Reynolds numbers. Changing the turbulence model will demonstrate its sensitivity to the eddy viscosity shape and to the friction coefficient.

12-9

Set up the problem first for a value of **Re** = 30,000. Choose properties for any fluid; since this is a constant-property problem, the choice will not matter. The integration should be carried out from **x** = 0 to **x** > hydrodynamic entry length, so let $L/D_h \approx 75$ to assure hydrodynamically fully-developed flow. As with the pipe flow, the large L/D_h is because the centerline velocity will require much longer to develop than the wall friction. If you choose air at room temperature, then a channel width of 10 cm and a length of 15 m should be sufficient. For the **Re** = 150,000 calculation, use the same channel width. We suggest you use a turbulence model that is a composite of a mixing-length for the near-wall flow and an eddy viscosity model for the outer part of the flow. The procedure suggested in the textbook works quite well.

Most of the features of Problem 12-7 apply to this problem. Plotting the ratio u_{cl}/V shows a continual acceleration of the centerline velocity as the velocity profile changes shape. Note the change is significantly less than for laminar flow. It is worthwhile to plot a set of normalized velocity profiles, **u/V**, over $0 \le y/h \le 1$ to show how the initially flat profile changes (**h** is the channel half-width). The centerline ratio will approach about 1.22 for the low **Re** and 1.18 for the higher **Re**. Carry out the analysis leading to an equation similar to (12-10), then, using your computational results, evaluate the "computational experimental" exponent for the power-law form of the profile. You can also assume a logarithmic relationship similar to Eq. (12-6), but without the radius effect, and plotting the velocity profile in wall coordinates will permit comparison of the log-region part of the profile.

One very obvious feature is the rapid development of **$c_f/2$** to its fully-developed value (about 15-20 hydraulic diameters for both **Re**). Recall that for parallel planes, the hydraulic diameter Reynolds number must be used. There exists several choices of fully-developed friction coefficient equations to evaluate the approach. The circular pipe Eq. (12-14) is quite accurate, although, as with pipe flow, the comparison is quite sensitive to the details of the mixing-length/eddy viscosity turbulence model. When experimental uncertainty is considered, perhaps the choice of equation is not too important. If possible, evaluate the eddy diffusivity at the fully-developed state, and compare to Figure 12-1 and Eq. (12-5) for both Reynolds numbers. Changing the turbulence model will demonstrate its sensitivity to the eddy viscosity shape and to the friction coefficient.

Chapter 13

13-1

$$t^+ = 278.3 + 2.195 \ln y^+ \qquad \text{for } Pr = 100$$

$$St = \frac{0.0287 Re_x^{-0.2}}{46.4 Re_x^{-0.1} + 0.9}$$

These results are very insensitive to whether the numerical part of the integration is terminated at $y+ = 100$, or at other such values as 80, or 150. It is of further interest to investigate the value of "a" that would result if an equation of the following type is used to correlate the results:

$$St\, Pr^a = 0.0287 Re_x^{-0.2}$$

13-2

At vanishing small Prandtl number the thermal boundary layer becomes very much thicker than the momentum boundary layer. In that case the velocity through the entire thermal boundary layer becomes essentially constant, the eddy diffusivity is zero, and the applicable solution is Eq. (10-11).

13-3

A solution based on the energy integral equation and a cubic parabola temperature profile is:

$$St = 0.165 Re_x^{-0.4} Pr^{2/3} [1 - \xi/x]^{-1/3}$$

(assuming ξ/x is near **1.00**)

A virtually identical result is obtained by solution of the energy differential equation using laminar boundary layer similarity methods.

13-4

$$St\, Pr^{0.5} = 0.0125 Re_{\Delta_2}^{-0.25} \left(\frac{Pr^{0.4} Re_{\Delta_2}}{Re_{\delta_2}} \right)^{5/36}$$

$$St\, Pr^{0.5} = 0.0142 Re_{\Delta_2}^{-0.25}$$

13-5

$$\mathbf{St\,Pr^{0.44}} = 0.0287\,\mathbf{Re_x^{-0.2}}\left(\frac{\Delta_2}{\delta_2}\right)^{-1/9}$$

13-6

At the throat:

$$\mathbf{St} = 0.287(\sin\theta)^{0.2}\,\mathbf{Pr}^{-0.4}\,\mathbf{Re_T}^{-0.2}$$

where θ is the angle between the wall and the axis of symmetry. $\mathbf{Re_T}$ is a Reynolds number based on throat diameter.

13-7

$$\mathbf{t_0(x)} = \mathbf{t_\infty} + 7\left(\frac{\mathbf{x}}{1.8}\right)^{0.2}\frac{\mathbf{I(0,1)}}{\mathbf{I(0,1/3)}};\quad 0 \le \mathbf{x} \le 0.6$$

$$\mathbf{t_0(x)} = \mathbf{t_\infty} + 7\left(\frac{\mathbf{x}}{1.8}\right)^{0.2}\frac{\mathbf{I(0,0.6/x)}}{\mathbf{I(0,1/3)}};\quad 0.6 \le \mathbf{x} \le 1.8$$

$$\mathbf{I(a,b)} = \int_{\mathbf{a}}^{\mathbf{b}}\left[1-\eta^{0.9}\right]^{-8/9}d\eta$$

If we let **u** = **1-$\eta^{0.9}$**, then,

$$\mathbf{I(a,b)} = \frac{10}{9}\int_{\mathbf{a}}^{\mathbf{b}}\mathbf{u}^{-8/9}(1-\mathbf{u})^{1/9}\,\mathbf{du}$$

This integral is of the form of an incomplete Beta Function, see Appendix C. The peak temperature is at **x** = **0.6** is about **59**°C. At **x** = **36** cm the temperature has dropped to about **9** °C. A heat flux of about 8600 W/m^2 is obtained.

It may be more convenient to solve the original integral numerically by computer.

13-8

The procedure for this problem is the same as for 13-7, an application of Eq. (13-33). This time the integral must be broken into four parts, corresponding to the four surface segments. The solution for the first two is the same as for the previous problem with the integration limits suitably modified.

13-9

$$\eta = 32.6\,\mathbf{Pr}^{0.4}\,\mathbf{Re_x^{0.2}}\,\mathbf{M}\left[\mathbf{P}\left(1-\left(\frac{\xi}{\mathbf{x}}\right)^{9/10}\right)-\mathbf{P}\left(1-\left(\frac{\xi+\mathbf{h}}{\mathbf{x}}\right)^{9/10}\right)\right]$$

where $$P(r) = \frac{\beta_r\left(\frac{1}{9}, \frac{10}{9}\right)}{\beta_1\left(\frac{1}{9}, \frac{10}{9}\right)}$$

x,ξ measured from the leading edge.
ξ is distance to beginning of strip heater.

13-10

This is an extension to the previous problem. A meaningful result can be obtained if the heat flux in the region downstream from the heater is set at about 10% of the heater value, and then at 5%. The question is whether or not a heat transfer coefficient based on $(t_0 - t_w)$ is at least approximately independent of $(t_0 - t_w)$.

13-11

For the 60 cm wide by 60 cm in the flow direction cooler, $Q \approx 2500$ W
For the 1.2 cm wide by 30 cm in the flow direction cooler, $Q \approx 2700$ W

13-12

For $Re_x < Re_{x,t}$, $St = 0.417\, Re_x^{-0.5}$

For $Re_x > Re_{x,t}$, $St = 0.033\left[Re_x - Re_{x,t} + 41.7\, Re_{x,t}^{0.625}\right]^{-0.2}$

13-13

If it is assumed that transition occurs at $Re_{\delta_2} = 300$, the following results are obtained:

θ degrees from stagnation point	h W/(m²K)
0	3.44
10	3.42
20	3.37
30	3.27
40	3.15
50	2.98
----transition----	
60	11.6
70	10.99
80	10.58
90	10.07

13-14

A reasonable way to handle the nose region is to use the potential flow solution for flow over a sphere:

$$u_\infty = (3/2)V\sin(x/R_0)$$

Then at the point where $\mathbf{u_\infty = V}$, let $\mathbf{u_\infty = V}$ over the remainder of the nose and along the cylindrical body. If it is assumed that transition to a turbulent boundary layer takes place when $\mathbf{Re_{\delta_2} = 307}$, transition will occur at **x = 0.4 m (38.2°)**.

The following table lists the heat transfer coefficients calculated as a function of distance from the stagnation point. Also listed are the values of **h** which would be calculated if it were assumed that the system were a simple flat plate with constant free-stream velocity, and that the turbulent boundary layer originated at the stagnation point:

x, m		h, W/(m2K)		h_{simple} plate
0		53.0		
0.1		52.8		
0.2		52.0		
0.3		50.7		
0.4		48.9	(transition)	
0.438		227.0	($\mathbf{u_\infty = 60}$)	
0.5		185.7		
0.6		163.8		
0.7		152.1		
0.8		143.7		
0.9425	(cyl. section)	134.0		116.34
1.943		106.6		100.7
2.943		96.1		92.6
3.943		89.7		87.4
4.943		85.3		83.5

13-15

The temperature and velocity profile origin should be shifted about **0.018** cm below the nominal origin. Then,

$$t^+ = 0.165 \ln y^+ - 0.348$$

This should then be compared with Eq.(13-46) using $\mathbf{Re_k = 124}$, $\mathbf{\kappa = 0.41}$, $\mathbf{Pr_t = 0.9}$. a quite good fit will be obtained.

13-16

x, m Dist. from leading edge	t_0,K convective	t_0,K transpiration	t_0,K film
0	800	800	-
0.04	682	586	290
0.08	674	572	533
0.12	669	564	623
0.16	666	557	664
0.20	663	552	689
0.24	600	548	706
0.28	658	544	717
0.30	657	543	726

13-17

Set up the problem for a constant free stream velocity of 10 m/s and a free stream temperature of 300 K (this is a duplicate of Problem 11-10). Let the properties be air at 1 atm, constant, and let the plate length be 10 m. Let the surface temperature be 295 K. Start the calculations at about 1.23 m. This should provide a momentum thickness Reynolds number range from 2,000 to 10,000. The initial profiles can be constructed as outlined in the problem statement. Analysis of the problem can include development of both the temperature profiles and the Stanton number

For the air temperature profiles, consider $\mathbf{Re}_{\Delta 2} = 4000$ and confirm the viscous sublayer, Figure 13-9, and the law-of-the-wall, Eq. (13-15), an well as the general shape of the profile in $\mathbf{t}^+$ versus $ln\ \mathbf{y}^+$ coordinates, as depicted in Figure 13-9. Plot the profile using a power-law form, similar to that leading up to Eq. (13-23) to determine what the power law is for this profile. Be sure to use a thermal boundary layer thickness that is reasonable, such as the location where $(\mathbf{t_o} - \mathbf{t})/(\mathbf{t_o} - \mathbf{t_\infty}) = 0.99$. Evaluate the profile to determine Δ_2 from Eq. (6-14). Pay particular attention of the uncertainty in evaluation of the enthalpy thickness.

If possible, plot the non-dimensional heat flux profile at $\mathbf{Re}_{\Delta 2} = 4000$ to confirm the Couette-flow part of the profile, Eq. (13-11).

Carry out the profile analyses at **x**-locations corresponding to $\mathbf{Re}_{\delta 2} = 6000$, 8000, and 10000 to confirm the inner-region similarity ($\mathbf{t}^+$ versus $ln\ \mathbf{y}^+$).

Plot the ln **St** distribution versus $ln\ \mathbf{Re}_{\Delta 2}$ to confirm Eq. (13-19) and versus $ln\ \mathbf{Re_x}$ to confirm Eqs. (13-17) and (13-18). You can also determine **St** by use of the energy integral equation, (6-24). Be careful of accuracy in creating the enthalpy thickness derivative and its uncertainty.

If possible, modify the turbulence model (**Pr_t**) to see what effect it has on the Stanton number results.

13-18

Set up the problem for a constant free stream velocity of 46 m/s and a free stream temperature of 265 K (see Problem 13-7). Let the properties be air at 1 atm, constant, and let the plate length be 1.8 m. Let the surface heat flux be 8600 W/m^2 over the first 0.6 m, followed by an adiabatic wall for the remaining part of the plate. Start the calculations at about 0.001 m, using a virtual origin that yields an **x**-Reynolds number of about 56,000, this should provide a momentum thickness Reynolds number of about 1500 at the start of the calculations. The initial profiles can be constructed as outlined in the problem statement, but pay particular attention to the thermal profile. Estimate the wall temperature by calculating a turbulent heat transfer coefficient and using the given wall heat flux. The shape of the thermal profile can be assumed the same as that obtained for a constant wall temperature turbulent temperature profile.

The results can be compared to the analysis of Problem 13-7. Notice how much more rapidly the turbulent boundary layer responds to a change in surface boundary condition compared to the laminar boundary layer (see, for example, Problem 10-20). By about 30 cm after the heating stops, the surface temperature has significantly fallen. This can be evaluated in terms of boundary layer thickness.

13-19

Set up the problem for a constant free stream velocity of 61 m/s and a free stream temperature of 269 K (see Problem 13-11). Let the properties be air at 71 kPa, constant, and let the plate length be 1.5 m. Let the surface temperature be equal to the free stream temperature over the first 0.9 m, followed by an isothermal section at 327 K for the remaining part of the plate. Start the calculations at about 0.001 m, using a virtual origin that yields an **x**-Reynolds number of about 50,000, this should provide a momentum thickness Reynolds number of about 1400 at the start of the calculations. The initial velocity profile can be constructed as outlined in the problem statement, and the thermal profile will be flat at the free stream temperature. You will need to integrate the heat flux distribution (by graph is ok) to obtain the total heat transfer per unit width. For the 0.6 m heating, the result should be about 4200 W/m, and for the 0.3 m heating, the result should be about 2300 W/m. Comparison to Problem 13-11 shows agreement with the problem solution, and it shows the high initial heat transfer at the step decays quite fast, in contrast to laminar decay, making the design of a shorter, wider heat transfer surface acceptable.

Chapter 14

14-1

Let **b** = plate spacing. Then $b^+ = b\sqrt{\tau_0/\rho}/\nu = bV\sqrt{C_f/2}/2$. Use a three-layer model with a laminar sublayer extending to **y+ = 13.2** . Under the assumptions suggested ε_M/ν becomes constant at **y+ = 0.394b+.** Then the temperature profile in the three regions is:

Sublayer - $t^+ = Pr\, y^+$

Log profile region - $$t^+ = t^+_{y^+=13.2} + \int_{13.2}^{y^+} \frac{dy^+}{\left(\frac{b^+}{2} - y^{+2}\right)}$$

b+ is determined from the results of Prob. 12-1, i.e.,

$$c_f/2 = 0.04Re^{-0.25} = 0.00274$$
$$b^+ = 1309$$

The center-line is at **y+ = b+/2 = 665.**
The value of **t+** at the center-line is **54.16.**

14-2

$$t^+ = \int_0^{y^+} \left\{ \left[\frac{-1}{\left(1 - \frac{y^+}{r_0^+}\right)\left(\frac{\varepsilon_H}{\nu} + \frac{1}{Pr}\right)} \right] \left[\frac{4}{Re} \int_0^{y^+} \left(u^+ + \frac{S r_0}{2 r_0^+} \left\{ u^+ - \frac{Re}{2 r_0^+} \right\} \right) \left(1 - \frac{y^+}{r_0^+} \right) dy^+ - 1 \right] \right\} dy^+$$

where $\dot{q}_0''$ is positive <u>into</u> the fluid and

$$r_0^+ = y^+_{y=r_0} = \frac{r_0 V\sqrt{c_f/2}}{\nu} = \frac{Re\sqrt{c_f/2}}{2}$$

If a 2 - layer model is used, for **0 < y+ < 13.2** (laminar sublayer), **u+ = y+,** and -

$$t^+ = Pr\left\{ y^+ - \frac{2y^{+3}}{3Re} + \frac{S r_0}{2\dot{q}_0''}\left(\frac{y^{+2}}{r_0^+} - \frac{2y^{+3}}{3Re} \right) \right\}$$

For **13.2 < y+ < r_0+** a closed form solution is obtainable if we assume $u^+ = 8.6 y^{+1/7}$.

Numerical integration using mixing-length theory, Eq. (11-25), $\varepsilon_M / \nu = 0.4 r_o^+ / 6$ in the central region, and Eq.(13-7) for $\mathbf{Pr_t}$, leads to the following results:

Pr	**Re**	$\dfrac{\mathbf{S r_o}}{2\dot{\mathbf{q}}_o''}$	**Nu**
4.0	50000	0	255.1
4.0	50000	-1	259.3
4.0	50000	+1	251.1

14-3

One side heated:

Pr = 10 **Nu = 680**
Pr = 0.01 **Nu = 6.70**

Both sides heated:

Pr = 10 **Nu = 712**
Pr = 0.01 **Nu = 11.96**

At **Pr = 10** the heat transfer resistance is concentrated very near the walls, while at **Pr = 0.01** it is distributed over the entire flow cross-section.

14-4

a) **33.2** W/(m^2 K) b) **44.2** (If laminar, then **22.2**)
c) Laminar, **35.5** d) **15732** e)**29605**
f) **54188** g) **1374** h) **1374**

14-5

x, m	t_m °C	t_w °C
0	10	10
0.5	20	140
1.0	40	156
1.5	56	171
1.8	65	180

14-6

ϕ	t_w °C
0	700
$\pi/2$	633
π	566
$3\pi/2$	633
2π	700

14-7

The results will differ a bit depending upon the calculation method used. The following are based on numerical integration using mixing-length theory.

(For an eigenvalue solution, the results of Prob. 9-10 are directly applicable with a simple and quite obvious modification.)

x, m	$(t_w - t_m)/(t_{w,max} - t_m)$	
	Pr = 0.7	Pr = 0.01
1.25	1.00	1.00
1.27	0.13	0.34
1.29	0.11	0.28
1.31	0.09	0.23
1.45	0.045	0.085
1.53	0.026	0.051

14-8

A finite difference solution of the applicable differential equations yields the following results for local **h**. A solution based on the data in Table 14-7 and superposition theory yields values of h somewhat higher, but otherwise the trends are similar. [Note: Problems involving use of Table 14 may yield somewhat different numerical results from those given for the problems to follow because of newer data in Table 14-7.]

x, m	h, W/m^2K
.758	123
.81	114
.88	110
1.01	106
1.26	103
1.81	102
1.96	102
2.03	110
2.08	108
2.18	105
2.31	104
2.50	102
3.23	102

14-9

Using the series solution functions of Table 14-7, the following results are obtained:

		Pr = 0.01	Pr = 0.7
Nu,	constant temperature	5.85	117.5
Nu,	linearly varying temperature	7.46	120.5

14-10

x, m	$\dot{q}_o''$, W/m²	t_m °C	t_i °C	t_o °C
0	80452	275	286	274
.5	130174	276	293	274
1.0	175029	279	301	277
1.5	210626	281	309	278
2.0	233481	284	315	281
2.5	241356	288	319	285
3.0	233481	291	321	288
3.5	210626	294	321	291
4.0	175028	296	319	293
4.5	130173	299	316	298
5.0	80452	300	311	299

14-11

The entry-length functions necessary to solve this problems precisely, taking into consideration axial variation of surface heat flux, are not available. If it is solved the same way as Prob. 14-10, the following results are obtained:

x, m	$\dot{q}_o''$, W/m²	t_m °C	t_i °C	t_o °C
0	23285	330	332	332
.5	37676	331	335	334
1.0	50658	334	339	337
1.5	60961	336	342	340
2.0	67576	339	346	344
2.5	69855	343	350	347
3.0	67576	346	353	350
4.0	50658	351	357	355
4.5	37676	354	357	356
5.0	23285	355	357	357

The effects of axial heat flux variation can be estimated by analyzing flow in an "equivalent" circular tube, which might be defined as having the same flow area and velocity. Thus **D = 0.0433**. Then Eq. (9-50) and the functions in Table 14-7 can be used.

14-12

The solution in this problem is based on the data in Table 14-3, i.e.,

$$Nu_{ii} = 79.8 \qquad \theta_i^* = 0.338$$

$$Nu_{oo} = 64.3 \qquad \theta_o^* = 0.055$$

together with Eqs. (9-28) and (9-29). However, since neither surface temperature is initially known, the heat transfer rate from the outer tube to the helium cannot be immediately calculated. Thus an iterative procedure is required. To start, assume that all the heat is transferred to the helium from the inner surface. This leads to an inner surface temperature of **235.5** °C and an outer surface temperature of **197.6** °C. A first estimate of the radiation exchange, and the outer surface convection, can now be made using these temperatures, and this leads to a second estimate of the heat fluxes and surface temperatures.

14-13

Set up the problem first for a value of **Re** = 30,000. Choose properties appropriate to each of the three Prandtl numbers to carry out constant-property calculations. Let the pipe diameter be 20 cm. The integration should be carried out from **x** = 0 to **x** > the thermal entry length, so let $\mathbf{L/D_h} \approx 75$ to assure thermally fully-developed flow for **Pr** = 0.01 and 0.7 and let $\mathbf{L/D_h} \approx 100$ for **Pr** = 100. This large $\mathbf{L/D_h}$ is because the heat transfer is **Pr** dependent, and like the hydrodynamics, the thermal profile adjustment lags the wall heat transfer. For the **Re** = 150,000 calculation, use the same radius. We suggest you use a turbulence model that is a composite of a mixing-length for the near-wall flow and an eddy viscosity model for the outer part of the flow (p. 248 of the 3rd Ed), and the textbook's variable turbulent Prandtl model for the **Pr** = 0.01 and 0.7; use a constant **Prt** of 0.85 for **Pr** = 100. Create a heat flux boundary condition using a suitable heat flux. For the liquid metal **Pr**, a wall heat flux of 5000 W/m^2 provides a 2 percent change in wall-to-mean temperature ratio for the lower **Re**, and a lower percentage change for the higher Re. For the air **Pr** , a wall heat flux of 300 W/m^2 provides a 3 percent change in wall-to-mean temperature ratio for the lower **Re**, and 100 W/m^2 provides a similar change for the higher Re. For the oil **Pr** , a wall heat flux of 100,000 W/m^2 provides a 3 percent change in wall-to-mean temperature ratio for the lower **Re**, and 300,000 W/m^2 provides a similar change for the higher **Re**.

For the liquid metal, there is a noticeable change in thermal entry length with **Re**, with the lower **Re** requiring about 15-20 $\mathbf{D_h}$ to develop, and about 30-35 $\mathbf{D_h}$ at the higher **Re**, reflecting the combined molecular and turbulent influence. The temperature profile development follows the Nusselt number development. For the air, 15-20 $\mathbf{D_h}$ is required for the two **Re**, similar to the hydrodynamic entry length. This pretty much follows Figure 14-8, although that figure is for an unheated starting length, whereas this problem is a combined entry length. One would expect the combined effect to produce a slightly higher Nusselt number ratio, than if only the thermal field were developing. Note that the temperature profile development somewhat lags the Nusselt number development. For the oil, 5-10 $\mathbf{D_h}$ is required for the two **Re**, longer than the hydrodynamic entry length. Again, the temperature profile development somewhat lags the Nusselt number development. Although this problem is a combined entry problem, the ideas depicted in Figures 14-6 and 14-7 can be confirmed.

14-14

Set up the problem first for a value of **Re** = 30,000. The fluid is air at **Pr** = 0.7. Let the pipe diameter be 20 cm. The integration should be carried out from **x** = 0 to **x** > the thermal entry length, so let $\mathbf{L/D_h} \approx 75$ to assure thermally fully-developed flow. This large $\mathbf{L/D_h}$ is because the heat transfer is because, like the hydrodynamics, the thermal profile adjustment lags the wall heat transfer. For the **Re** = 150,000 calculation, use the same radius. We suggest you use a turbulence model that is a composite of a mixing-length for the near-wall flow and an eddy viscosity model for the outer part of the flow (p. 248 of the 3rd Ed), and the textbook's variable turbulent Prandtl model. Create a wall temperature boundary condition using a wall temperature that is about 10 K above (or below) the inlet temperature. The direction of the heat transfer does not matter for these constant property calculations.

For air , a thermal entry length of about 15-20 $\mathbf{D_h}$ is required for the two **Re**, similar to the hydrodynamic entry length. The temperature profile development somewhat lags the Nusselt number development. The ratio of fully-thermally developed Nusselt numbers for this constant wall temperature boundary condition is about 3.5 percent lower, similar to that suggested by comparison of Eq. (14-12) with Eq. (14-7).

14-15

Set up the problem first for a value of **Re** = 30,000. Choose properties appropriate to each of the three Prandtl numbers to carry out constant-property calculations. Let the plate spacing be 10 cm. The integration should be carried out from **x** = 0 to **x** > thermal entry length, so let $\mathbf{L/D_h} \approx 75$ to assure thermally fully-developed flow. This large $\mathbf{L/D_h}$ is because the heat transfer is **Pr** dependent, and like the hydrodynamics, the thermal profile adjustment lags the wall heat transfer. For the **Re** = 150,000 calculation, use the same plate spacing. We suggest you use a turbulence model that is a composite of a mixing-length for the near-wall flow and an eddy viscosity model for the outer part of the flow (p. 248 of the 3rd Ed), and the textbook's variable turbulent Prandtl model for the **Pr** = 0.01 and 0.7; use a constant **Prt** of 0.85 for **Pr** = 100. Create a heat flux boundary condition using a suitable heat flux. For the liquid metal **Pr**, a wall heat flux of 5000 W/m^2 provides a 2 percent change in wall-to-mean temperature ratio for the lower **Re**, and a lower percentage change for the higher **Re**. For the air **Pr** , a wall heat flux of 300 W/m^2 provides a 3 percent change in wall-to-mean temperature ratio for the lower **Re**, and 100 W/m^2 provides a similar change for the higher **Re**. For the oil **Pr** , a wall heat flux of 100,000 W/m^2 provides a 3 percent change in wall-to-mean temperature ratio for the lower **Re**, and 300,000 W/m^2 provides a similar change for the higher **Re**.

Analysis of the flow for these three Pr follows the discussion of Problem 14-13. For Pr = 0.01, a comparison with theory can not be made because of the laminar influence. That is, you can not use Eq. (14-10) along with a change to hydraulic diameter Reynolds number. For **Pr** = 0.7 and 100, comparisons can be made using, say, Eq. (14-8) to monitor the approach to thermally fully-developed flow. The variation of Nusselt number with $\mathbf{x/D_h}$ is similar to that for a pipe.

14-16

Set up the problem for a value of **Re** = 19,445 based on the given Problem 14-5 parameters . The pipe diameter is 1.25 cm and the length is 1.8 m. We suggest you use a turbulence model that is a composite of a mixing-length for the near-wall flow and an eddy viscosity model for the outer part of the flow (p. 248 of the 3rd Ed), and the textbook's variable turbulent Prandtl model. Create a heat flux boundary condition based on the energy balance calculation for Problem 14-5, **q"** = 205,882 W/m^2.

Results of the calculation show the heat transfer coefficient reaches a fully-developed value of about 1,700 $W/(m^2\ K)$ at a $\mathbf{x/D_h} \approx 30$, and the pipe length is $\mathbf{L/D_h} = 144$. Actually, **h** is near its fully-developed value by $\mathbf{x/D_h} \approx 10$, which helps to verify the analysis required for Problem 14-5, namely a constant **h**. Note that for this wall heat flux boundary condition, the wall temperature rises from an entry value of 10 deg C to a value of 186 deg C at exit. This is a mean-to-wall ratio of 0.736. For fluids of this range of **Pr**, a variable property calculation will make a significant difference (although there is not sufficient information for such a calculation).

14-17

Set up the problem for a value of **Re** = 100,000 and for the two **Pr** given in Problem 14-7. The pipe diameter is 2 cm; let the pipe length be 1.53 m. We suggest you use a turbulence model that is a composite of a mixing-length for the near-wall flow and an eddy viscosity model for the outer part of the flow (p. 248 of the 3rd Ed), and the textbook's variable turbulent Prandtl model. Create a heat flux boundary condition for the **Pr** = 0.01 case of **q"** = 500,00 W/m^2, and for the **Pr** = 0.7 case use **q"** = 10,00 W/m^2. The heating is carried out for the first 1.25 m, and then an adiabatic condition is required.

Because of the constant heat flux boundary condition, the bulk mean temperature rises linearly until the adiabatic section, and then is constant from that point. The wall temperature rises in response to the heating. For both cases the heat transfer coefficient and friction coefficient are fully-developed at the start of the adiabatic section ($\mathbf{x/D_h} = 50$). The decay of the bulk mean temperature, in non-dimensional form $\mathbf{(t_o - t_m)/(t_{o,max} - t_m)}$, can be plotted versus the length scale $\mathbf{x/D_h}$, and can be compared with the results of Problem 14-7. Here $\mathbf{t_{o,max}}$ is the wall temperature at the start of the adiabatic section. The air temperature at the start of the cooling section is much less that for the liquid metal, as expected from the relative magnitudes of the heat transfer coefficients, about 200 $W/(m^2\ K)$ compared to 10,000 $W/(m^2\ K)$, but the air temperature, as a percentage of its value at the start of the adiabatic section, decays much more rapidly, reflecting the lower turbulent Prandtl number, and hence higher eddy diffusivity for heat transfer.

14-18

Set up the problem for a value of **Re** = 50,000 and the fluid air, as given in Problem 14-8. The pipe diameter is 2.5 cm and the pipe length is 3.25 m. We suggest you use a turbulence model that is a composite of a mixing-length for the near-wall flow and an eddy viscosity model for the outer part of the flow (p. 248 of the 3rd Ed), and the textbook's variable turbulent Prandtl model. Create a wall temperature boundary condition that is equal to the inlet fluid temperature (21 deg C) for the first 0.75 m, step-increased to 40 deg C for the next 1.25 m, and then step-increased again to 50 deg C for the final 1.25 m.

The adiabatic section (the first 0.75 m) corresponds to $x/D_h = 30$, enough flow distance to assure a hydrodynamically fully-developed flow. At the start of heating, the Nusselt number will become large, decaying as an unheated starting length problem over the next $x/D_h \approx 10\text{-}15$, becoming thermally fully-developed. At $x/D_h = 80$, the next step increase in wall temperature once again increases the Nusselt number, followed by a decay. In both heating intervals, the heat transfer coefficient is high only for 10 of the 50 x/D_h, allowing the assumption of constant **h** to be a fairly good one.

14-19

Set up the problem for a value of **Re** = 50,000 and for the two **Pr** given in Problem 14-9. Let the pipe diameter be 2 cm and the pipe length be 15 m. We suggest you use a turbulence model that is a composite of a mixing-length for the near-wall flow and an eddy viscosity model for the outer part of the flow (p. 248 of the 3rd Ed), and the textbook's variable turbulent Prandtl model. Create a linearly-varying wall temperature boundary condition that is equal to an assumed inlet fluid temperature, say 300 K, and ramps up linearly to 350 K at the exit, at $L/D_h = 75$.

For **Pr** = 0.01, you will find the Nusselt number continually changing, with the ratio **Nu/Nu,fd** quite high over the first 10 x/D_h, and remaining at a slowly decaying ratio of about 1.4 - 1.3 over the remaining 75 x/D_h. This is in marked contrast to the air flow, which responds more quickly to the changing thermal boundary condition. This is similar to the numerical Problem 14-17. The thermal entry effect is mostly gone by At $x/D_h \approx 20\text{-}30$ and the ratio **Nu/Nu,fd** is near 1.0 for the remainder of the flow. Again, this is a turbulent Prandtl number effect, reflecting the presence of the laminar conductivity contribution to the flow. You would also expect a **Re** effect for the low **Pr** case.

Chapter 15

15-1

h = 56 W/(m²K) $\quad c_f$ **= 0.027**

Referring to Eqs. (15-3) and (15-4), **n = -0.34** and **m = 0.48.**

15-2

For this problem the student should select some suitable values for pressure, free-stream temperature, free-stream velocity, and distance x along the surface. Then t0 and h should be evaluated for several values of surface temperature, both above and below free-stream temperature, using Eqs (8-13), (8-12), and (10-13), first by evaluating all properties at free-stream temperature and using the exponents fro Table 15-1, and second by evaluating all properties according to Eq. (15-15). The result should be quite comparable although not necessarily identical.

15-3

This problem is the same except that the equations under investigation are (11-23) and (13-18), and the temperature ratio exponents suggested are **-0.4** for $(T_0/T_\infty)<1$ and **0.0** for $(T_0/T_\infty)>1$. For this problem the comparison of methods may not be so favorable but there exists virtually no experimental data upon which to base a judgement.

15-3

Chapter 16

Note: Many of the results for this chapter have been obtained using mixing-length theory and a finite-difference program (a predecessor of STAN7). Somewhat different results may be obtained if integral methods are used.

16-1 $\dot{q}_0'' = 31$ kW/m²

16-2 $\dot{q}_0'' = 11$ kW

16-3

Numerical integration is required. At the large values of **h** considerable error is introduced unless care is taken and a fairly sophisticated integration scheme is employed.

16-4

x, cm (axial distance)	$\dot{q}_0''$ kW/m²
1.94	119
6.07	141
8.14	774
10.2	889
12.26	1282
13.5	1872
14.33	1246
15.15	802
16.39	637
18.45	434
20.51	302
22.58	217
24.64	185
26.72	128
28.75	164
30.81	159
34.95	77

16-5

The following results have been obtained assuming an initial laminar boundary layer with transition at $\mathbf{Re}_{\delta_2}$ = **300**. (At the throat it makes little difference whether an initial laminar boundary layer is assumed, or not.) The assumed wall conductivity is **17.3** W/(mK).

z, cm (axial distance)	**h** W/(m²K)	$\mathbf{T_w}$ K
0	-	-
1.94	284	455
6.07	337	475
8.14	1848	713

16-5 con't

z, cm (axial distance)	h W/(m²K)	T_w K
10.2	2126	729
12.26	3162	761
13.5	4278	812
14.33	3821	713
15.15	2401	687
16.39	1704	684
18.45	1188	636
20.51	877	585
22.58	689	545
24.64	569	520
26.72	466	
28.75	414	508
30.81	368	505
34.95	272	425

The following is a set of experimental data points obtained for the nozzle under study. Experimental uncertainty is about ± **15** percent.

z, cm (axial distance)	h W/(m²K)
6.71	1143
12.27	3169
12.97	4470
13.51	3820
14.54	3492
14.74	2707
15.24	2375
15.65	2069
18.16	1241
20.72	901
23.39	808
28.34	478
31.02	384
33.49	333

16-6

z, cm (axial distance)	h W/(m²K)	$\dot{m}''$ kg/(s m²)
0	-	-
1.94	234	0.535
6.07	526	1.198
8.14	688	1.568
10.2	935	2.128
12.26	1573	3.484

16-6 con't

z, cm (axial distance)	h W/(m²K)	$\dot{m}''$ kg/(s m²)
13.5	2195	5.272
14.33	2147	3.975
15.15	1377	2.775
16.39	950	2.216
18.45	678	1.583
20.51	515	1.157
22.58	440	0.914
24.64	345	0.747
26.72	298	0.556
28.75	246	0.655
30.81	218	0.631
34.95	182	0.353

Total transpiration flow rate = **0.13** kg/s.

16-7

Altitude m	Re_x	T_{aw} K	h W/(m²K)	$\dot{q}_o''$ kW/m²
0	0	288	0	0
1000	10e06	283	66.6	-2.68
4500	53e06	287	164	-5.92
10000	63e06	336	172	2.22
20000	12.9e06	678	65.0	23.09
30000	2.0e06	1245	20.4	18.82
40000	0.6e06	2006	10.4	17.5
--- Transition to laminar---				
50000	97900	2867	1.41	3.59
60000	17800	3729	0.71	2.42

16-8

To fit the specifications of the problem an initial velocity must be assumed. Using **V = 10** m/s as the velocity at zero altitude, and a density of **7849** kg/m³ for the metal skin, the following results are obtained:

Altitude m	Flight time, s	Surface temp, K
0	0	288
1000	35.8	288
4500	63.1	288
10000	78.6	293
20000	92.3	323
30000	100.3	354

16-8 con't

Altitude m	Flight time, s	Surface temp, K
40000	106.1	367
50000	110.5	373
60000	114.1	374

16.9

	$\dot{q}_0''$, kW/m^2 Upper surface	$\dot{q}_0''$, kW/m^2 Lower surface
Stag point	1016	1016
Distance from stag point, cm		
0.2	431	295
0.5	250	255
1.0	193	125
---Transition---		
2.0	352	250
3.0	420	241
5.0	346	204
6.0	306	-

16-10

There is no "correct" answer for this problem. The results of Prob. 16-9 are to be used for a design study, but the types of internal flow passages that could be used are quite varied. The problem is to get a high enough heat transfer coefficient on the inside surface so as to pull the surface temperature down. Internal fins are a possible answer.

16-11

For the upper (convex) surface only, the following results are obtained:

	$\dot{m}''$, kg/(s m^2)
Stag point	2.1
Distance from stag point, cm	
0.2	1.66
0.5	1.56
1.0	1.67
2.0	1.88
3.0	1.75
5.0	1.54
6.0	1.48

16-12

z, cm (axial)	h, W/(m²K)	$\dot{q}_o''$, W/m²	Re_{Δ_2}
0	∞	∞	0
1	1604	1.77e06105	
2	1263	1.39e06	161
3	1156	1.27e06	216
4	1134	1.25e06	274
5	1161	1.27e06	341
6	1232	1.35e06	420
8	1500	1.63e06	654
----transition----			
10	7233	7.78e06	1187
12 (throat)	7742	8.03e06	1938
14	5849	5.70e06	2777
16	3767	3.56e06	3752
18	2509	2.31e06	4858
22	1372	1.23e06	6712

16-13

The results of this problem depend almost completely on the ingenuity of the student in devising a suitable cooling system. It will be found to be extremely difficult to hold the surface uniformly at **100** °C, but the objective of the problem is to see what <u>can</u> be done. Local boiling may have to be accepted. Although surface temperature will probably not be constant, it will be sufficiently accurate to use values of **h** from Prob. 16-12.

16-14

x, m (along surface)	h, W/m2K	$\dot{m}''$, kg/(s m²)	B
0.0138	1458	0.380	1.35
0.0275	1412	0.368	1.35
0.0413	1480	0.386	1.35
0.055	1612	0.420	1.35
0.0688	1805	0.470	1.35
0.0825	2075	0.540	1.35
0.110	2970	0.771	1.35
0.137	4196	1.07	1.33
0.157	4620	1.15	1.29
0.175	3582	0.849	1.23
0.198	2352	0.545	1.20
0.221	1591	0.461	1.18
0.267	886	0.197	1.15

Up to this point the total coolant required is **0.037** kg/s, less than one percent of the thru-flow, which is **5.28** kg/s. A turbulent boundary layer has been assumed from **x** = **0**.

16-15

For the **60** cm by **60** cm cooler, $\dot{q} \approx$ **3600** W. For the **1.2** by **30** cm cooler, $\dot{q} \approx$ **3900** W. No cooling is possible when **V** reaches **360** m/s (**800** MPH).

Chapter 17

17-1

The development parallels that on pages 64-65, except that $u_\infty = 0$ and a body force acts in the x-direction.

$$-\tau_0 = \frac{d}{dx}\int_0^y \rho u^2 dy + y\frac{dP}{dx} + \int_0^y \rho g dy$$

Now, from hydrostatics, $dP/dx = -\rho_\infty g$, and let $\rho_\infty - \rho = \rho\beta(t - t_\infty)$. See pages 397-398 for explanation.

$$-\tau_0 = \frac{d}{dx}\int_0^y \rho u^2 dy - g\beta\int_0^y \rho(t - t_\infty)dy$$

17-2

For the first case; $(t_0 - t_\infty) \propto x^{1/5}$

$$\overline{(t_0 - t_\infty)} = \frac{1}{L}\int_0^L (t_0 - t_\infty)dx = \frac{5}{6}(t_0 - t_\infty)_{x=L}$$

From the definition $Nu_x = C \cdot Gr_x^{1/x}$, at x = L,

$$\frac{\dot{q}_0''}{\Delta t_L} \cdot \frac{L}{k} = C \cdot \left(\frac{g\beta}{\nu^2}\Delta t_L L^3\right)^{1/4}$$

Substitute , $\Delta t_L = \frac{6}{5}\overline{\Delta t}$, find

$$\overline{Nu} = (6/5)^{(5/4)} \cdot C \cdot Gr_L^{1/4} = 1.256 \cdot C \cdot Gr_L^{1/4}$$

For the second case, $(t_0 - t_\infty)_{L/2} = (1/2)^{1/5}(t_0 - t_\infty)_L$. (Recall that $\Delta t \propto x^{1/5}$). Substitute into the Nusselt number expression. Find $\overline{Nu} = \left(2^{1/5}\right)^{5/4} \cdot C \cdot Gr_L^{1/4}$. Use Table 17-2 for C and compare with Table 17-1, modified by Eq.(17-19). Sample results:

Pr = 10 $\quad Nu_T$ = 1.10 $\quad \overline{Nu_Q} = 1.17 \quad \overline{\overline{Nu}}|_Q$ = 1.11

17-3

For v >> u, the applicable equation is (17-21) with F' = 0 and F = constant = F(0).

$$\theta'' + Pr(n+3)F\theta' = 0$$

with $\theta(0)=1$ and $\theta(\infty)=0$. Integrate to obtain the following:

$$\theta = \exp[-(n+3)\ Pr\ F(0)\ \eta]$$

The Nusselt number becomes:

$$Nu_x = (n+3) Pr \frac{F(0)}{\sqrt{2}} Gr_x^{1/4}$$

Comparison with Table 17-3 shows agreement within 2-1/2 % for $F(0) \geq 0.8$.

17-4

Modify the problem statement: Let the ribbon be **1** m wide by **3** m high. Its energy dissipation is **0.5** W/cm^2 to air at **25**^{o}C.

The solution must be iterated because t_0 is unknown. First guess for $(t_0 - t_\infty)$ would be based on $h \approx 8$ W/(m^{2}K) [order of magnitude for free convection].
After 3 - 4 iterations find $Gr_L^x \approx 3x10^{13}$. Thus the ribbon is mostly in turbulent flow. Find $\overline{Nu} = Nu_L \approx 380$, $h = 6.1$ W/(m^{2}K), and $t_0 = 820$ ^{o}C.

17-5

This problem is developed in some detail in Ref. 16. Substitute the profiles and expressions for **u** an δ into the integral equations (17-24, 17-25) to obtain two algebraic equations, each of which contain **x** to the exponential powers **m** and **n**. The algebraic equations must be valid for all **x**; hence equate the exponents. Find **m = 1/2** and **n = 7/10**. Substitute these back into the algebraic equations and solve for C_u and C_δ:

$$C_u = 0.0689 \nu C_\delta^{-5} Pr^{-8/3}$$

$$C_\delta = 0.00338 \frac{\nu^2}{g\beta(t_0 - t_\infty)} \cdot (1 + 0.494 Pr^{2/3}) Pr^{-16/3}$$

Introduce the Grashof number definition, with the result:

$$\frac{\delta}{x} = 0.0565 Gr^{-1/10} Pr^{-8/15} [1 + 0.494 Pr^{2/3}]^{1/10}$$

17-6

For this problem $Gr_L = 2.3 \times 10^6$. For horizontal, upward facing surfaces, Eq.(17-33) give **Nu = 19.4**. For inclined surfaces, the **+45°** orientation is out of the recommended range for using **g cosθ**, but for this low Gr_L, it is probably OK. Find $\overline{Nu} = 17.1$. For a vertical surface, $\overline{Nu} = 18.6$. Note, all values within 12%.

17-7

For the horizontal case, $\mathbf{Gr_D = 4.1 \times 10^{-7}}$. Using Eq. (17-34) (laminar), and Eq. (17-35) for correction, find $\mathbf{Nu = 0.37}$. For the vertical orientation, Eq.(17-36) is not satisfied. Thus Eq, (17-14) must be corrected by (17-35). Find $\overline{\mathbf{Nu}} = 2.3$. Note that Eq.(17-14) must be multiplied by **4/3** before using (17-35).

Chapter 18

18-1

The differential equation for steady heat transfer along a fin is:

$$\frac{d\hat{t}}{dx^2} - \left(\frac{hp}{kA}\right) t = 0$$

where **h** is the heat transfer coefficient along the surface (assumed constant), **p** is the periphery of the fin (assumed constant), **k** is the thermal conductivity, and **A** is the cross-section area of the fin (constant).

The boundary conditions usually relevant to a cooling fin are:

$$t = t_0 \text{ at } x = 0$$
$$\frac{dt}{dx} = 0 \text{ at } x = \ell$$

The resulting solution for the temperature distribution is then:

$$\frac{t}{t_0} = \frac{\cosh m(\ell - x)}{\cosh m\ell}$$

where $m = \sqrt{\frac{hp}{kA}}$

From the temperature gradient at **x = 0** the heat transfer rate at the base of the fin can be evaluated:

$$\dot{q}_{x=0} = kAt_0 m \tanh m\ell$$

Were the entire fin at t_0, (conductivity infinite), the heat transfer rate at the base would be:

$$\dot{q}_\infty = hp\ell t_0$$

The fin "heat transfer effectiveness is then defined:

$$\eta = \frac{\dot{q}_0}{\dot{q}_\infty} = \frac{\tanh m\ell}{m\ell}$$

18-2

Let $\Delta t = t_h - t_c$ and $C_h < C_c$. Then the following differential equation can be developed from an energy balance on each fluid side and a heat transfer rate equation connecting the two sides:

$$\frac{d(\Delta t)}{\left(\frac{1}{C_c}-\frac{1}{C_h}\right)} = U\,dA\,\Delta t$$

Integrating from one end of the heat exchanger to the other,

$$\ln\frac{\Delta t_2}{\Delta t_1} = \frac{AU}{C_c} - \frac{AU}{C_h}$$

Let $NTU_c = AU/C_c$ and $NTU_h = AU/C_h$. Then:

$$\frac{\Delta t_2}{\Delta t_1} = e^{NTU_c - NTU_h} = \frac{t_{h_1} - t_{c_1}}{t_{h_2} - t_{c_2}}$$

By definition (for $C_h < C_c$) -

$$\varepsilon = \frac{C_c\left(t_{c_1} - t_{c_2}\right)}{C_h\left(t_{h_1} - t_{c_2}\right)} = \frac{C_h\left(t_{h_1} - t_{h_2}\right)}{C_h\left(t_{h_1} - t_{c_2}\right)}$$

After combining equation and solving for ε,

$$\varepsilon = \frac{1 - e^{-(NTU_h - NTU_c)}}{1 - \frac{C_h}{C_c} e^{-(NTU_h - NTU_c)}}$$

Now since $C_h < C_c$, $C_h = C_{min}$ and $C_c = C_{max}$, the result can be generalized by substituting:

$$NTU_h = NTU_{max}$$

$$NTU_c = NTU_{min} = AU/C_{max} = (C_{min}/C_{max})NTU_{max}$$

which will then yield Eq. (18-8). Eq.(18-9) is obtained by applying L'Hospital's Rule to Eq. (18-8) and setting $C_{min}/C_{max} = 1$, or it can be obtained by re-deriving the differential equation for this special case.

18-3

The method of analysis is described on page 432. For the hot fluid the energy balance for each cell would be:

$$\frac{C_h}{N} t_{h_k} = \frac{C_h}{N} t_{h_{k+1}} + \dot{q}_i$$

For the cold fluid,

$$\dot{q}_i + \frac{C_c}{M} t_{c_i} = \frac{C_c}{M} t_{c_{i+1}}$$

The rate equation connecting the two flow passages is:

$$\dot{q}_{ik} = U\left(\frac{A}{NM}\right)\left[\left(\frac{t_{h_k} + t_{h_{k+1}}}{2}\right) - \left(\frac{t_{c_i} + t_{c_{i+1}}}{2}\right)\right]$$

Starting at one corner of the heat exchanger where the entering temperatures are known, it is then possible to calculate the outlet temperatures of each cell, moving along each row and then stepping down to the next row. Once the outlet temperatures are established along the outlet faces of the heat exchanger, the mean outlet temperatures can be computed, and then the effectiveness, ε, from the definition of effectiveness. The whole process can be easily programmed and will involve two nested loops.

18-4

The trick to using a spreadsheet for this kind of calculation is to use 2 or more spreadsheet cells to represent one cell of the heat exchanger. A spreadsheet cell is needed for each variable that needs to be calculated. For this problem the minimum that needs to be evaluated for each cell in the heat exchanger is the cell outlet temperature for each of the two fluids. Thus at least 2 cells are need. More cells can be used if desired. For example it might be desired to know the rate of heat transfer in each heat exchanger cell, in which case 3 spreadsheet cells would be used. In complex problems where fluid properties are variable and where pressure drop must be calculated, as many as 15 cells might be used to represent one heat exchanger cell. If the equations become very complex the use of multiple cells is often helpful.

For this problem the equations to be solved for each heat exchanger cell are the same as in the previous problem. Once the spreadsheet cells for one heat exchanger cell have been built, the COPY command can be used to replicate the cells to the entire heat exchanger, and of course the spreadsheet program automatically changes the cell references as necessary.

There then remains the question of boundary conditions. The spread sheet cells bounding the physical heat exchanger can be used for boundary values. The problem specifies a two-pass heat exchanger and this is easily handled by appropriately changing the cell references in the cells where fluid enters. The mean temperature of a fluid coming out of a heat exchanger, or leaving one pass, is obtained by summing the temperatures and dividing by the number of flow passages.

Once the spreadsheet has been set up, advantage should be taken of a spreadsheet's ability to re-calculate. Although in simple cases it may be possible to obtain a solution in one pass of calculations, more often iteration must be used, but this involves no more than pressing a single key on the keyboard as many times as necessary to obtain convergence.

The student should be encouraged to look into more complex problems and suggest how he might handle such things as variable properties, multiple fluid heat exchangers, etc. Also note that the scheme can be easily used for a simple cross-flow heat exchanger.

Chapter 19

19-1

The air outlet temperature will be about 61 °C.
The water outlet temperature will be about 75 °C.
The pressure drop on the air side will be 1060 Pa .
The pressure drop on the water side will be 21 Pa.

19-2

The dimensions of the heat exchanger will depend upon which particular heat exchanger surfaces are chosen for the two fluid sides. The following are the dimensions if the surfaces chosen are:

Air Side: #12.00T
CO_2 Side: #1/8-15.61

(This choice of surfaces is purely arbitrary and does not necessarily represent the smallest design.)

The dimensions will be as follows:

Air Side flow length = 0.535 m
CO_2 Side flow length = 0.651 m
Non-flow length = 0.341 m
Volume = 0.119 m^3

19-3

If the same surfaces are used as in the previous problem, but a two-pass cross-counterflow arrangement is used, the dimensions will be:

Air Side flow length (one pass) = 0.46 m
CO_2 Side dimension (2-passes) = 0.28 m
Non-flow length = 0.734 m
Volume = 0.095 m^3

19-4

Again, the result will depend upon the surfaces chosen. The following results are based on the use of Surface #1/8-15.61 on both sides, and aluminum construction:

Fluid 1 flow length = 0.237 m
Fluid 2 flow length = 0.237 m

Non-flow length = 0.436 m
Percent pressure drop (both sides the same) = 0.16

This result yields a rectangular duct. If a nearer to square duct is desired it would be well to investigate one of the other surfaces. It is also worth noting that a rotating periodic flow heat exchanger is very practicable for this application and has been used. If a compact matrix is employed a considerably smaller unit will result.

This problem requires a considerable amount of iteration. It is very time-consuming to work by hand but lends itself well to computer calculations. [See note on the availability of Computer Software, page. xii in the text.]

19-5

The heat exchanger dimensions, and other characteristics, will depend heavily upon the chosen pair of surfaces. The following results are obtained using:

High pressure air side: Surface #12.000T
Low pressure combustion products side: Surface #1/8-15.61

The percent pressure drop was distributed 3 percent to each side.

The results are:

Heat exchanger dimension in air flow direction (2-passes) = 1.072 m
Dimension in combustion products direction (1-pass) = 0.385 m
Non-flow dimension = 1.121 m
Volume = 0.462 m3
Weight = 524 kg.

Of particular note in this design is that the fin effectiveness on the air side is only 0.72, and on the combustion gas side is only 0.60. This suggests that the fins are too thin and/or too long. Note that the weight is less than 1.0 kg/HP so that the overall weight is favorable relative to competing systems, but a better choice of surfaces can reduce this figure considerably. [See also what can be accomplished with a rotating periodic flow design in the next problem].

19-6

Assuming equal distribution of the percent pressure drop (3% to each side), the results are:

High pressure air frontal area = 0.09 m^2
Combustion gas frontal area = 1.42 m^2
Total frontal area (circular cross-section) = 1.50 m^2

Flow length = 0.1469 m
Weight = 92.8 kg

Here the great advantage of the rotating heat exchanger can be seen.

Chapter 20

20-1

The following relations are needed:

From Eq.(20-26) -

$$\dot{m}'' = \frac{\sum_j [\gamma_j(\partial m_j/\partial y)]_0 + [\Gamma c \partial t/dy]_0}{i_0 - i_T}$$

$$i_T = i_L - \dot{q}''_L/\dot{m}''$$

$$\sum_j [\gamma_j(\partial m_j/\partial y)]_0 i_{j,0} = [\gamma_1(\partial m_1/\partial y)]_0 i_{1,0} + [\gamma_2(\partial m_2/\partial y)]_0 i_{2,0}$$

$$m_1 + m_2 = 0 \qquad i_L = i_{2,0}$$

$$\frac{\partial m_1}{\partial y} + \frac{\partial m_2}{\partial y} = 0 \qquad i_0 = (m_1 i_1)_0 + (m_2 i_2)_0$$

$$\gamma_1 = \gamma_2$$

$$\dot{m}'' = \frac{[\gamma_j(\partial m_j/\partial y)]_0}{m_{j,0} - m_{j,T}} = \frac{\gamma_1(\partial m_1/\partial y)_0}{m_1}$$

Combining, we obtain $\dot{q}''_0 = \dot{q}''_L$

20-2

If **Lewis No. = 1**, then, $\dot{m}'' = g\frac{i_\infty - i_0}{i_0 - i_T} = \frac{g(i_\infty - i_0)}{i_0 - i_L + \dot{q}''_L/\dot{m}''}$

Solve for $\dot{q}''_L$. Then in addition,

$$\dot{m}'' = \frac{(m_{2,\infty} - m_{2,0})}{(m_{2,0} - m_{2,T})} = \frac{-g\, m_{2,0}}{m_{2,0} - 1} \; ; \qquad m_{2,0} = \frac{1}{1 + g/\dot{m}''}$$

Introducing the relations from the previous problem for the enthalpies we will obtain:

$$\dot{q}_L'' = \dot{q}_0'' = g(i_{1,\infty} - i_{1,0})$$

If c_1 = constant, then,

$$\dot{q}_0'' = g c_1 (t_\infty - t_0)$$

From which, $\dfrac{g}{G_\infty} = St$

20-3

If **i** is a "conserved" quantity,

$$\dot{m}'' = g\frac{(i_\infty - i_0)}{(i_0 - i_T)} = g\frac{(i_\infty - i_0)}{\left(i_0 - i_L + \dot{q}_L''/\dot{m}''\right)}$$

$$\dot{m}'' i_0 - \dot{m}'' i_L + \dot{q}_L'' = g(i_\infty - i_0)$$

Let $\dot{m}''$ = **0**. Also $\dot{q}_L'' = \dot{q}_0''$. Then,

$$\dot{q}_0'' = g(i_\infty - i_0)$$

20-4

Refer to pages 258-259 in the text. Let there be a concentration gradient for some component "**j**". Then the rate of transfer of than component in the direction of the gradient would be:

$$G_{diff,j} = C\sqrt{\overline{v'^2}}\,\rho\,\delta\overline{m_j}$$

The procedure then follows in complete analogy with that for turbulent heat transfer. Let ε_{diff} be the eddy diffusivity for mass diffusion. Then,

$$\delta\overline{m_j} \approx \ell\frac{d\overline{m_j}}{dy}\,; \qquad \frac{G_{y,diff,j}}{\rho} = \varepsilon_{diff}\frac{d\overline{m_j}}{dy}\,; \qquad \varepsilon_{diff} = \varepsilon_M$$

$$Sc_T = \frac{\varepsilon_M}{\varepsilon_{diff}} = 1.00$$

20-5

See the development starting on page 373. This leads to Eq.(16-5) for a laminar boundary layer, or Eq.(16-7) for a turbulent boundary layer. Let **Pr = 1.00**, or $\mathbf{Pr_{eff}}$ = **1.00.** In either case the resulting equation is the same as Eq. (20-21).

Chapter 21

21-1

The equation which must be solve is:

$$\frac{g}{g^*} = \frac{1}{B\gamma_{m_{CO_2}\to 0}} \int_0^{m_{CO_2,0}} \gamma_{CO_2} \frac{d\,m_{CO_2}}{1-m_{CO_2}}$$

where $B = \dfrac{m_{CO_2,0}}{1-m_{CO_2,0}}$

Using data from Table A-19, and integrating numerically:

$m_{CO_2,0}$	B	g/g*
0	0	1.0
0.2	0.25	0.925
0.5	1.00	0.769
0.8	4.00	0.488

21-2

For this case take γ_{CO_2} outside of the integral. Then,

$$\frac{g}{g^*} = \frac{\ln(1+B)}{B}$$

The following modification of this result fits the results of 21-1 reasonably well:

$$\frac{g}{g^*}\left(\frac{M_\infty}{M_0}\right)^a = \frac{\ln(1+B)}{B}$$

where a = 0.6
M is molecular weight

21-3

The equation to be solved is:

$$G_r \frac{dP}{dr} - \frac{\lambda}{r^2}\frac{d}{dr}\left(r^2 \frac{dP}{dr}\right) = 0$$

$$g = \frac{\lambda \ln(1+B)}{r_0 B}; \quad \frac{g}{g^*} = \frac{\ln(1+B)}{B}$$

21-4

$m_{C_6H_6,0}$	B	g/g*
0	0	1.00
0.2	0.25	0.959
0.5	1.00	0.850
0.8	4.00	0.607

Again, these results correlate well with the constant property results using the following equation:

$$\frac{g}{g^*}\left(\frac{M_\infty}{M_0}\right)^a = \frac{\ln(1+B)}{B}$$

21-5

Axisymmetric stagnation point:

B	g/g*
0	1.0
0.950	0.633
3.56	0.342

Two-dimensional stagnation point:

B	g/g*
0	1.0
1.195	0.592
4.97	0.294

These should be compared to the results in Fig. 21-1 for $\mu/\lambda = 0.7$.

21-6

$$\dot{m}'' = 0.031 \quad \text{kg/(sm}^2\text{)}$$

$$m_{C_6H_6,0} = 0.47$$

Chapter 22

22-1

In O_2, **B = 15.8**

In air, **B = 3.65**

$$\dot{m}''_{in\,O_2}/\dot{m}''_{in\,air} = 1.82$$

22-2

t_0 = -29 °C (Based on data from *Handbook of Chemistry and Physics*)

22-3

t_0 = 54 °C

$\dot{m}''$ = 0.0013 kg/(sm²)

22-4

t_0 = 15.2 °C

$\dot{m}''$ = 0.00019 kg/(sm²)

(It is difficult to evaluate **B** and $\dot{m}''$ precisely for this problem.)

22-5

$\dot{m}''$ = 0.047 kg/s (Total from both sides.)
Boundary layer assumed turbulent over entire surface.
Re_x = 750,000
B = 0.32 , g = 3.24

22-6

With reaction, **B = 0.98,** $(m_{H_2})_0$ = **0.48**
With no reaction, **B = 0.79,** $(m_{H_2})_0$ **= 0.44**

For the first case a "simple" chemical reaction is assumed, together with Lewis No. = 1.0. It is assumed that the "flame" is detached so that $(m_{H_2})_0$ = **0.** There will be inert substances at the surface (N_2 and combustion products), since only **0.48** is H_2. Note also that $(m_{O_2})_\infty$ = **0.232** (atmospheric air.) In the second case **B** is computed from the energy equation alone, but the combustion case requires both the energy and the simple chemical reaction mass diffusion equations. In both cases, i_∞ = **13,805,000** J/kg.

22-7

From Prob. 16-9, **h = 4700** W/(m²K). Then $g^* = h/c =$ **4.05** kg/(sm2), t_0 = **124.5** °C, **B = 0.30**, **g = 3.54** kg/(sm2), $\dot{m}''$ = **1.06** kg/(sm²) (**0.099** kg/(ft²), **0.22** lbm/(sft²). Note that Eq.(22-4) must be used to determine $m_{H_2O,0}$ because the local pressure is the stagnation pressure, **51600** Pa. Table A-20 only goes to **100** °C, so vapor pressure data must be obtained form other sources.

22-8

Because of the extensive calculations involved, it is suggested that this problem be developed on a computer. Pressure must first be determined along the blade surface using the tabulated values of **r** and **T**, Prob. 11-8, then at each point m_{H_2O} as a function of **T** must be calculated from vapor pressure data, after which local values of t_0 and **B** can be computed as in Prob. 22-7. Values of g^* can be determined from values of **h** calculated for Prob. 16-9.

22-9

Molecular weight of mixture = **21.5** kg/kmol
Specific heat of mixture (c_p) = **1546** J/(kgK)
Ratio of specific heats, **k = 1.33**
Mass velocity at throat, **G = 2030** kg/(sm²)
Viscosity of mixture at throat, **m = 7.67e-05** Pa s
Pressure and temperature at throat, **P = 1836000** Pa
T = 2807 K

g^*/G at the throat can be determined from the results of Prob. 13-6,

$$g^*/G = 0.0268\,Pr^{-0.4}\,Re_T^{-0.2}$$

$$g^* = 3.17 \text{ kg/(sm}^2)$$

The chemical system and the possible reactions are essentially the same as on Pages 529-530. **B = 0.215**, **g = 2.87** kg/(sm2), $\dot{m}''$=**0.617** kg/(sm²). If the graphite density is **2500** kg/m3, the time for a **10** percent increase in throat diameter is **40.5** seconds, assuming that **g** does not change.

22-10

$\dot{q}''_0$ = **657** kW/m²
T_0 = **1850** K

For graphite density = **2200** kg/m³, the rate of ablation is **0.006** inches/minute (**0.015** cm/min).

22-11

$\dot{m}''$ = 0.335 g

If the ice is treated as a sphere of radius D, we can obtain:

$$g = 2.0 \frac{\Gamma}{D} \frac{\ln(1+B)}{B}$$

where **G = k/c**

22-12

$\dot{m}$ = 0.000274 kg/s

$t_0 = t_L$ = 26 °C

B = 0.0113

$m_{H_2O,0}$ = 0.00125

22-13

$$\dot{m}'' = \frac{\Gamma \ln(1+B)}{r_0}$$

Time for complete consumption of the droplet is:

$$\theta = \frac{\rho R_0^2}{2\Gamma \ln(1+B)}$$

where R_0 is the <u>initial</u> radius.

B = 9.45

G = k/c = 6.12e-05 Ns/(m2)

q = 1.92e-05 s

22-14

The equations on page 519 may be used for enthalpy. It is only necessary to reduce the T-state enthalpy by **333** kJ/kg, the "latent heat" of fusion. assuming that the bulk of the snow is at **-23** °C,

$$i_T = 4.2(-23) - 333 = -430 \text{ kJ/kg}$$

$$t_0 = t_L = 5\ °C$$

The snow is evidently melting before evaporating..

$m_{H2O,0} = 0.0051$, $B = 0.0048$, $\dot{m}'' = 2.88e-05$ kg/(sm²)
If the specific heat of the snow is about **0.1**, the snow is evaporating at a rate of **1** mm/hr.)

22-15

The mixture temperature is **38** °C. With **20** percent air present (the air ejector is provided to hold this percentage low), condensation at this point in the condenser can only occur if **B < 0**. This requires that $t_0 = t_L < 35.6$ °C. The cooling water and the tube surface must then be less than **35.6** °C. One can then write:

$$q''_L = \frac{t_0 - t_{wall}}{R}$$

A **0.1** mm liquid film would give **R = 1.63e-04**. If there were no air in the system this would be essentially the only resistance. With the Lewis No. = **1.0** assumption, the effective additional resistance can now be investigated by assuming various values of t_{wall} and evaluating $\dot{m}''$. Then compare with what the condensation would be were $t_0 = 38$ °C, **R = 1.63e-04**, and $\dot{m}'' = \dot{q}''/L$, where **L = 2409** kJ/kg, the latent heat of vaporization.

22-16

g, kg/(sm2)	$\dot{m}''$, kg/(sm²)	B	t_0, °C
0.0188	0.0104	0.552	80
0.0362	0.0101	0.279	70
0.1109	0.00963	0.0868	50
0.1978	0.00971	0.049	40
0.2720	0.00998	0.0367	35
0.590	0.0119	0.0202	25
2.67	0.0286	0.0107	15

22-17

g, kg/(sm2)	$\dot{m}''$, kg/(sm²)	B	t_0, °C
0.0063	0.00349	0.552	80
0.0162	0.00452	0.279	70
0.0745	0.00647	0.0868	50
0.155	0.00761	0.0491	40
0.229	0.0837	0.0367	35
0.562	0.0114	0.0202	25
2.84	0.0304	0.0107	15

22-18

$\dot{m}_{steam} / \dot{m}_{air}$	t_0	B	$\dot{m}''$
0	6.0	0.00578	0.00577 g*
0.01	17.8	0.00275	0.00275 g*
0.02	25.9	0.00142	0.00142 g*

These results suggest that mixing steam at this temperature with the dry air will reduce drying rate despite higher system temperatures.

22-19

$\dot{m}''$ = **-0.00193** kg/(sm²)
$\dot{q}''_L$ = **4.85** kW/m²
t_0 = **49.5** °C   B = **-0.0274**

A condensate film thickness of **0.01** mm would increase resistance by **22** percent.

22-20

$\dot{m}''$ = **-4.17** kg/(sm2) (condensation rate)
$\dot{q}''_L$ = **11.58** kW/m²
B = **-0.29**
g* = **0.0122**
$t_0 = t_L$ **29.1** °C
Re_x = **43626** (laminar boundary layer)

22-21

This problem can be approached at least two ways. In the first the Lewis No. = **1.00** approximation can be used and constant properties assumed. In the second the variable properties solution shown on Fig. 22-5 can be used. In the first, Eq. 21-10 is used for g. The differences provide a good way to assess the accuracy of the more approximate procedure.

Re_x = 1,509,393 (turbulent) **g^* = 0.148**

Air injection:

t_0	$\dot{m}''$, kg/(sm²) const. prop. Eq. 21-10 for g	$\dot{m}''$, kg/(sm²) variable prop. Fig. 22-5 for g/g*	B
800	0.0526	0.0549	0.426
600	0.0992	0.106	0.954
400	0.164	0.180	2.039

22-21 con't

	t_o	$\dot{m}''$, kg/(sm²) const. prop.	$\dot{m}''$, kg/(sm²) variable prop.	B
Helium injection:				
	800	0.0121	0.130	0.085
	600	0.0259	0.0283	0.191
	400	0.0507	0.0477	0.408

For the case of air injection, the experimental data on Fig. 13-19 would suggest that Eq. 21-10 is more accurate than Fig. 22-5 for constant properties and **Pr = 0.72.**

22-22

A modification of the conserved property used on Page 530 can be employed:

$$P = n_c - \frac{3}{4} n_o = m_c - \frac{3}{11} m_{CO_2} - \frac{2}{3} m_{H_2O} - \frac{3}{4} m_{O_2}$$

First consider two cases where O_2 is added to the original mixture. (m_{N_2} is **0.4** in the original mixture.):

Case	m_{O_2}	m_{CO_2}	m_{H_2O}	m_{N_2}	B
1 (original)	0	0.4	0.2	0.4	0.243
2	0.10	0.36	0.18	0.36	0.293
3	0.20	0.32	0.16	0.32	0.344

Rate of ablation increases.

Next consider two cases where N_2 is added to the original mixture:

Case	m_{O_2}	m_{CO_2}	m_{H_2O}	m_{N_2}	B
1 (original)	0	0.4	0.2	0.4	0.243
4	0	0.333	0.167	0.5	0.202
5	0	0.267	0.133	0.6	0.161

Rate of ablation decreases.

Finally it is of interest to let m_{CO_2} and m_{H_2O} go to **0.0**, and let m_{O_2} = **0.232**, m_{N_2} = **0.768** (atmospheric air). Then **B = 0.174**, the same as for the "simple chemical reaction".

- NOTES -

- NOTES -

- NOTES -

- NOTES -